Detection Technologies for Chemical Warfare Agents and Toxic Vapors

Detection Technologies for Chemical Warfare Agents and Toxic Vapors

Yin Sun and Kwok Y. Ong

CRC Press
Taylor & Francis Group
Boca Raton London New York

CRC Press is an imprint of the
Taylor & Francis Group, an **informa** business

CRC Press
Taylor & Francis Group
6000 Broken Sound Parkway NW, Suite 300
Boca Raton, FL 33487-2742

First issued in paperback 2019

ISBN-13: 978-1-56670-668-1 (hbk)
ISBN-13: 978-0-367-39402-8 (pbk)

Library of Congress Card Number 2004047815

Library of Congress Cataloging-in-Publication Data

Yin Sun, 1962-
Detection technologies for chemical warfare agents and toxic vapors / Yin Sun, Kwok Y. Ong.
p. cm.
Includes bibliographical references and index.
ISBN 1-56670-668-8 (alk. paper)
1. Chemical warfare. 2. Chemical agents (Munitions)--Measurement. 3. Gases, Asphyxiating and poisonous--Measurement. I. Ong, Kwok Y. II. Title.

UG447.S779 2004
623.4'592—dc22 2004047815

Visit the Taylor & Francis Web site at
http://www.taylorandfrancis.com

and the CRC Press Web site at
http://www.crcpress.com

Preface

Terrorists have used toxic chemicals and high-powered explosives on many occasions. Examples include bombing of the World Trade Center in 1993, deployment of the chemical agent sarin in the Tokyo subway system in 1995, and the bombing of the Oklahoma City Federal Building in the same year. The events of September 11, 2001, in New York City and Washington, D.C., changed the lives of people in the U.S. as well as in many other nations throughout the world.

Sarin is one of several toxic chemicals defined as weapons of mass destruction (WMDs). WMDs "are capable of a high order of destruction and/or of being used in such a manner as to destroy large numbers of people. Weapons of mass destruction can be high explosives or nuclear, biological, chemical, and radiological weapons, but exclude the means of transporting or propelling the weapon where such means is a separable and divisible part of the weapon. Among weapons of massive destruction, besides high explosives, are the chemical, biological, and nuclear weapons" (*Dictionary of Military and Associated Terms*, U.S. Department of Defense, Joint Publication, 2002). Use of WMDs, such as nuclear, biological, or chemical weapons by terrorists toward civilian targets, will cause large numbers of casualties. The recent anthrax episode in the U.S. is a prime example of how a biological agent could cause casualties and fear. The anthrax incidents together with the Tokyo subway incident, which killed more than ten people and injured thousands, shocked the world into awareness that even a small-scale deployment of these WMDs could cause chaos. The use of WMDs against civilian targets is no longer unlikely, nor is any country today spared from such terrorist acts. The potential use of chemical warfare agents (CWAs) and/or toxic industrial compounds (TICs) by terrorist groups has increased substantially in recent years.

CWAs are relatively inexpensive to manufacture. Release of these agents could be carried out surreptitiously, with the first awareness of exposure occurring when victims develop symptoms. The U.S. government has attempted to tighten controls on these compounds and necessary reagents for their syntheses. There are, however, many other compounds used in industries that are also quite toxic. These toxic industrial compounds or chemicals (TICs) are readily available in large quantities. Although they are less toxic compared to conventional CWAs, their availability in large quantities has made them attractive to terrorist groups.

Given that terrorists can synthesize CWAs or easily obtain large amounts of TICs, the threat of using CWAs or TICS at any time toward government or civilian targets is cause for great concern. It is not possible to predict when incidents may happen, regardless how much effort is spent in prevention. Therefore, fast and reliable equipment to detect the existence of these compounds in the air is critically needed to offer protection when such incidents occur. Authorities around the world are searching for effective rapid detection of airborne CWAs or TICs to permit orderly evacuation of exposed areas to minimize potential casualties.

We have been involved in testing and evaluation of CWA detection devices that include most of the existing point-sampling detectors for many years. It is hoped that this book will help the general public as well as decision makers in the public and private sectors to understand methodologies employed for the detection of toxic chemicals. This book provides pertinent physical, chemical, and toxicological information of the more dangerous compounds that can kill or cause severe injuries. We condense information about detection requirements, discuss U.S. government policies, explain various technologies used in detecting CWAs and TICs, and summarize the characteristics of many current detection devices. We have also included vapor generation techniques and analytical methodologies with recommendations on preferred apparatus and instrumentation for efficient laboratory evaluations of detection devices. We provide reasonably up-to-date information on toxic vapor detection with special emphasis on CWAs, which is also applicable to TICs.

We have attempted to provide as much useful information as possible on existing detection technologies. Time and space constraints prevented us from including all existing devices and methodologies; our discussion is then limited to the most common ones that we have encountered. Inclusion of these relatively common devices and methodologies does not mean that the U.S. government (or any other government) endorses them, or that we endorse them.

Acknowledgments: We could never have completed this book without the help of many people. We are especially grateful for the understanding and support of our families. Suggestions and advice from many reviewers are also greatly appreciated.

Authors

Yin Sun was born in XuZhou, JiangSu, China. He earned his B.S. degree at Nanjing University and his M.S. degree from the ChengDu University of Technology. He was awarded the Ph.D. degree by the University of Connecticut. Sun has worked in the areas of analytical chemistry, environmental chemistry, application chemistry, and analytical instrumentation for about 20 years. In the early stage of his career, he was interested in the geochemical behavior of noble metals, such as gold and silver. His research focused on finding valuable deposits using trace element distributions and combinations. Later, environmental issues, especially heavy metal pollution, became the focus of his research. His research centered on using radioactive isotopes as tracers for the behavior of heavy metals in estuaries. During this period, he developed new analytical techniques for the tracers he was using, and conducted many studies using various modern instruments.

Several years ago, Sun turned his attention to developing analytical methods for the detection and analysis of explosives and illegal drugs in various matrices. He then spent a number of years studying the detection of toxic industrial compounds and chemical warfare agents (CWAs). Sun has developed vapor generation systems and tested various techniques for the detection of TICs. Since CWAs are strictly controlled, he has spent much time in the U.S. Army Edgewood Chemical Biological Center (ECBC) laboratory. While working at the ECBC, his "second home," he met Kwok Y. Ong

Kwok Y. Ong received his B.S. degree in chemistry from California State University at Los Angeles. He was drafted into the U.S. Army immediately upon graduation and was assigned to the Edgewood Arsenal, headquarters of the Chemical Corps under the Science and Engineering Program. He remained at Edgewood as a civilian employee after discharge from the military until his retirement from the civil service in September 2001. He is currently associated with the EAI Corp., in which capacity he provides continued support to the government with special emphasis on CWA detection and detector evaluations.

Ong has extensive work experience with practically all point-sampling detectors, starting with thc U.S. military's M43 to the current ACADA, and including many evaluations of devices for the Domestic Preparedness program that were either developed in-house or abroad. He has patented vapor generation techniques and analytical methodologies that have become industry standards. Ong is a greatly respected expert in the field of toxic chemical detection. People refer to testing at the ECBC as in "Kwok's laboratory." His recommendations are considered authoritative.

Contents

Chapter 8 Infrared Spectroscopy

Chapter 9 Surface Acoustic Wave and Electrochemical Techniques

CHAPTER 1

Introduction

By definition, weapons of mass destruction (WMD) cause massive destruction and a large number of casualties. The use of a WMD involves deployment of chemical, biological, or radiological weapons. Chemical weapons have been used throughout history, and their development continues throughout the world. Recently, chemical weapons have been used against civilians, while abandoned chemical weapons cause casualties to innocents unless properly disposed of. Sensitive and reliable detectors are desperately needed to provide advance warning of chemical agent exposure to reduce potential casualties. Many governments strongly support research and development of technologies aimed at building improved detectors.

1.1 HISTORICAL OVERVIEW

Historically, humankind has used poisonous chemicals to disable or kill insects, fish, or other animals for various purposes. Chemical warfare agents are poisonous chemicals that can rapidly cause death or disability to the enemy. The deployment of chemical weapons is the use of poisonous compounds in time of war with the intention to kill or incapacitate large numbers of the enemy. In World War I, tear gases, phosgene, chlorine, mustard gases, and other respiratory impairment agents were used. For example, German soldiers deployed chlorine gas on April 22, 1915, resulting in the deaths of more than 5,000 Allied troops. The overall casualty toll from chemical weapons during WWI is estimated at 100,000 deaths and 900,000 injuries.

Many countries voted against the use of chemicals as weapons and signed the "Protocol for the Prohibition of the Use in War of Asphyxiating, Poisonous or Other Gas, and of Bacteriological Methods of Warfare" in 1925 at Geneva. Meanwhile, however, the development of poisonous chemicals to kill, incapacitate, or irritate enemy soldiers continued. During World War II, nerve agents such as tabun and sarin were developed and stockpiled by the Germans; tens of thousands of concentration camp victims were killed with chemical gases.

The Japanese also used chemical weapons during WWII. The Japanese Imperial Army injured and killed close to 100,000 people during the war using chemical and biological weapons. An estimated two million chemical warfare munitions and approximately 100 tons of toxic chemicals were abandoned in China alone when Japan surrendered. These abandoned chemical munitions continue to inflict casualties. As recently as August 4, 2003, mustard gas leaking from an abandoned Japanese chemical weapons plant in northeast China killed at least 1 civilian and injured 35 others. Abandoned chemical weapons in China have caused an estimated 2,000 deaths since WWII.

Development of chemical weapons continued after WWII. In 1952, scientists from the U.K. discovered a nerve agent more toxic than the G-agents, called VX. VX is a sulfonated organophosphorous compound that is substantially more toxic than the G series of nerve agents. In contrast to the G agents, VX is a persistent agent with extremely low volatility (10.5 mg/m^3 at 25°C). Because of its toxicity and persistency, VX is considered the most dangeous CWA. During the Vietnam War, chemicals used by the U.S. to defoliate vegetation are said to have caused deaths after contact with the chemicals. Other incidents of casualties caused by chemical weapons include the use of mustard and nerve agents during the war between Iraq and Iran in the 1980s. Iraqi soldiers used nerve agents against Iraqi civilians in 1988, resulting in the deaths of about 5,000 people. Chemical weapons have been used in recent terrorist attacks. The Sarin deployment by the Aum Shinrikyo in the Tokyo subway system in 1995 that killed more than ten people and injured thousands was a vivid example of the disastrous effects of even a small-scale release of chemical agents.

As history has demonstrated, the use of chemical warfare agents (CWAs) has caused a significant number of casualties. To minimize these numbers as a result of a chemical agent attack or accidental release of such chemicals, a general understanding of CWA behaviors by soldiers and the general public is extremely important. Knowledge will lessen the degree of anxiety that may lead to panic and ensuing catastrophe in the event that an incident occurs. Well-informed individuals would know the proper protection and evacuation procedures to minimize exposure and prevent the spread of contamination.

1.2 CHEMICAL WARFARE AGENTS

There are several classes of CWAs designed for different purposes with the intent to harass, disable, or kill people en mass. CWA characteristics follow:

- Toxicity
- Stability
- Can be easily made in mass production
- Can be disseminated in sufficient concentration in the field to produce desired effect

- Transportability
- Little or no corrosive action on storage containers
- Ability to minimize effectiveness of enemy's protective equipment
- Known mechanism of action, protection measures, and method of treatment
- Difficulty of detection before onset of physiological or psychological effects in targeted people
- Colorless, odorless, and nonirritating, yet toxic on exposure

Not all known CWAs possess all of the above characteristics, although the "ideal" agent would. Chemical warfare compounds are grouped according to their intended uses and effects on the human body. The best known lethal agents include nerve, blister, choking, and blood agents.

1.2.1 Nerve Agents

Nerve agents cause an increase in acetylcholine throughout the body. Acetylcholine is the substance that interferes with the functioning of the enzyme cholinesterase. Thus, nerve agents are also known as cholinesterase inhibitors or anticholinesterase agents. Acetylcholine plays the vital role of controlling the skeletal muscles, autonomic ganglia, and many structures of the central nervous system. Nerve agents produce symptoms such as respiratory difficulty, drooling and excessive sweating, nausea, vomiting, and cramps. Nerve agents are extremely toxic. Death can occur within minutes when a sufficient dosage enters the body through the respiratory system. Symptoms develop more slowly after skin exposure. Although a lethal dose may occur in 1 to 2 min of exposure, death may be delayed for 1 to 2 hr. Nerve agents include the G-agents (fluorine- or cyanide-containing organophosphates tabun [GA], sarin [GB], soman [GD], and cyclosarin [GF]) and V-agents (sulfur-containing organophosphorus compounds VX and Vx).

1.2.2 Blister Agents

Blister agents are used with the intention to injure or inflict casualties, often with the intent of slowing down troop movements. These agents affect the eyes and lungs of large numbers, who then require medical attention from other personnel. Exposure to high concentrations will cause eventual death. Mustard gas (HD) is one of the common blister agents that include the nitrogen mustards and Lewisite.

1.2.3 Choking Agents

These agents cause inflammation and swelling in the respiratory tract. The secretion of excess fluid in response to the irritation leads to coughing; when coughing becomes inadequate to the task of ridding the lungs of fluid, the person begins to choke. Death occurs when the person literally "drowns" in his own body fluid. Choking agents include phosgene (CG) and diphosgene (DP). Phosgene caused more than 80% of the CWA fatalities in WWI.

1.2.4 Blood Agents

Blood agents enter the body mainly via respiration. They prevent the normal use of oxygen by the blood cells and cause damage to tissues. These agents include hydrogen cyanide (AC), cyanogen chloride (CK), and arsine (SA).

1.2.5 Other Types of Agents

Incapacitating agents produce physiological or mental effects that may persist for several hours or days after exposure. Such effects hinder the enemy's effectiveness in battle, but do not seriously endanger their lives. Except in very high doses, an individual will recover from exposure to this type of agent.

Vomiting compounds (which, when heated, vaporize and condense to form aerosols) cause great discomfort to victims, thereby rendering them less effective in fighting. Included in this class of compounds are diphenylchlorarsine (DA), adamsite (DM), and diphenylcyanoarsine (DC). Excepting high-dosage cases, victims will recover in several hours.

Tear-producing compounds cause weeping and skin irritation. They cause transient injury and are widely used for riot control. Becasue of their fast acting and nonlethality, tear gases are commonly used in training of students to measure their ability in donning protective gears in the event of a CWA attack. Principal tearing compounds include chloroacetophenone (CN); chloroacetophenone in chloroform (CNC); chloroacetophenone, chloropicrin, and chloroform (CNS); chloroacetophenone, benzene, and carbon tetrachloride (CNB); bromobenzylcyanide (CA); and *O*-chlorobenzylidene malononitrile (CS).

Other types of less toxic CWAs are grouped according to their intended uses, such as riot control agents, training agents, smoke screen and signal chemicals, and defoliants. This book focuses on those chemical agents that adversely affect humans in particular. Most modern detection devices aim to detect nerve, blood, blister, and choking agents because of their high toxicity and lethality.

1.3 TOXIC INDUSTRIAL COMPOUNDS

Unlike CWAs that are manufactured explicitly to kill or incapacitate, thousands of chemicals used in modern industry run the gamut from mildly to extremely toxic. Some of these were considered CWAs in the past. For example, AC and CK were once classified as blood agents and phosgene was the choking agent CG. Because they are being used by manufacturers in many applications, they were removed from the more restrictive list of controlled CWAs to facilitate regular transportation in railway tanker cars or via tanker trucks over roadways. Thus, these compounds are now listed as highly toxic industrial compounds (TICs).

TICs are easily accessible in large quantities by potential terrorists. While detection of these compounds was not considered very important in early phases of developing CWA detection devices, they are now receiving much attention. Homeland Defense has added TICs to detector requirements.

1.4 CWA AND TIC DETECTION

Numerous methods, techniques, and instruments have been developed for the detection of CWAs and TICs. After the September 11 incident, aggressive efforts have focused on obtaining better detectors for diverse scenarios.

1.4.1 Historical Overview

During World War I, mustard gas and other chemicals such as chlorine gas were used on the battlefield. Detectors based on color changes resulting from reactions between decomposed mustard and appropriate reagents were developed. However, all detectors developed during the war, including the most common one of using trained soldiers to sniff the air for the characteristic garlicky odor of mustard gas, were not sensitive enough or fast enough to prevent casualties.

After WWI, efforts to build detectors that could rapidly detect chemical agents, especially mustard gas, led to development of standardized colorimetric detector kits, such as the M4 Mustard Agent Vapor Detector Kit, M5 Liquid Detector Paint, and M9 Chemical Agent Detector Kit. The M9 used a hand pump to sample air through a tube containing adsorbent and colorimetric reagent; detection was based on the change in color of the reagent.

Historically, developing detection techniques almost always lagged behind development of the chemical agents themselves. After the discovery of nerve agents by German scientists during World War II, the science of detection techniques faced even greater challenges because of these agents' high toxicity. The first fast-action field detector kit for nerve agents, the M9A2 Chemical Agent Detector Kit, was standardized in 1952.

Automatic CWA detectors developed in the United States in the 1960s and 1970s included the M43 and the M43A1 alarms, which were devoted to detect nerve agents. Technology for automatic detection of mustard vapor was not developed. Subsequently, the ability to detect mustard blister agents has become a necessity. Although not as lethal as nerve agents, HD is a carcinogen and the average person can generally detect its odor at very low concentrations. Because it is a carcinogen, there is no safe exposure level above the arbitrarily set allowable airborne exposure limit (AEL). The AEL of HD is currently set at 0.003mg/m^3 or less over an 8-hour exposure period. Much effort has been expended since the 1970s to advance HD detection. Most high-quality modern detectors are required to detect both nerve and blister agents.

1.4.2 Detection Requirements and Detector Development

While detection techniques for mustard gas and nerve agents are far from perfect at present, requirements to detect TICs as well pose even greater challenges to the developers. Because TICs are numerous with equally diverse properties, detecting them alongside blister and nerve agents is very complex. Minimum requirements for a high-quality CWA and TIC detector are most, if not all, nerve and blister agents, and some TIC compounds.

CWA and TIC detectors used in the field would be subject to diverse environmental conditions. The instrument needs to operate at a wide range of temperature and humidity levels. To provide adequate advance warning, detection sensitivity needs to be well below immediate danger to life and health (IDLH) concentration levels. This means that the combined dosage effect from the detectable concentration level and the time required for the detection must be sufficiently low to allow an individual sufficient time to protect herself once alerted. The detector should be able to detect these compounds with specificity. Detection must not be affected by coexisting substances in the atmosphere that could cause false-positive or false-negative responses. Of course, the ideal detector would use no consumables and work forever, and thus needs no additional supplies and provides noninterrupted protection. Unfortunately, the ideal detector does not exist now and will not exist in the future. Most sensors that have been mass-produced are designed for use in specific environments to detect one or two compounds.

Advances in analytical chemistry, microtechnologies, and computer software have made more techniques available for CWA and TIC detection. Detectors have become smaller, more sensitive, more reliable, and with more functions. Modern computer technologies have permitted more sophisticated data processing to enable fine-tuning of detection algorithms. Further miniaturization of detectors that are more sensitive, with lower false-alarm rates, and high tolerance to varied environment conditions summarizes the current focus of development.

The United States needs to improve and expand the uses of sensors in preventing terrorism and to minimize the impact should an incident occur. Besides point sampling devices, sensors to help provide sensitive and rapid detection and advance warning of toxic vapor at fixed sites such as subways, buildings, financial centers, and airports are of utmost importance. These sensors need to be operable around the clock. For example, sensors installed in the ventilation system could be coupled with a rapid shutdown procedure. Portable sensors to allow assessment from a remote or on-site point can be used to map the potential extent of the chemical cloud cover to aid authorities in organizing the movement of people. Current sensors have limited capabilities and must be improved.

To develop robust sensors, a multidisciplinary systems approach should be taken. Experimentalists, statisticians, engineers, and data analysts should collaborate from the beginning of a concept to the fielding of the final product. Statistically designed experimentation helps in reducing the need for exhaustive testing during development to produce a field-worthy sensor. Actual and potential interference must be identified and dealt with either through hardware design, multiple sensor types, multivariate techniques, or through sophisticated software development. Developing new, integrated multiple-source databases to create libraries for quick identification and to permit access to different methodologies is necessary. Manufacturers have already created many libraries. There is a need for consolidation to avoid redundancies. Researchers of diverse technologies need to collaborate and to share their expertise rather than limiting their applications to their own fields. A multisensor system is needed to provide the broadest detection capability possible.

Once developed, detectors and sensors must be thoroughly tested according to rigorous criteria. Sensor calibration and potential drifts in detection algorithm caused by operating environment conditions such as variations in humidity, temperature, and/or atmospheric pressure need to be addressed and corrected. As technologies evolve, more stringent requirements are being developed and applied to testing and evaluation of detection devices. Consequently, a number of different subsystems are needed to support development of these more sensitive instruments. Systems for reliable sample collection, sample processing, and presentation of chemical vapor to sensors are essential. Standardization of proven methodologies is needed. Systematic quality assurance for sensor evaluations can only be achieved through the use of standardized methodologies that have been proven successful. For example, mere testing for detection sensitivity using uncontrolled conditions has proven to be insufficient. Many devices perform differently under variable conditions. Temperature, relative humidity, moisture contents, and other substances in the atmosphere can affect detector performance. This is especially true during an incident. Many substances could influence the detection algorithm, resulting in false alarms in the absence of the CWAs (false positive) or their opposite in the presence of CWAs (false negative). Therefore, we must recognize that in any attempt to simulate real-world situations in the laboratory, many artifacts could arise. Care must be taken to recognize these possibilities and be receptive to make changes to eliminate all such artifacts.

Individual manufacturers may test newly developed detectors with CWA simulants and some TICs. Since CWAs are strictly controlled, their detectors cannot be tested with actual CWAs except in very few laboratories. The results from simulants cannot and should not be construed the same as with CWAs. In view of the many restrictions on the use of CWAs, the federal government offers the opportunity for developers to conduct laboratory, field, and wind tunnel tests through Test Service Agreements (TSA). A well-defined and very demanding set of test standards, including minimum detection levels, initiated by the Department of Defense, has been in place for the last 15 years. These standards include plans and methodologies for testing under conditions simulating real-world scenarios where possible. Evaluation criteria for detectors tested in government facilities follow:

- Accuracy of detected agent concentrations and dosages
- Consistency of exposure response times for repeated exposures at specific agent concentration levels
- Magnitudes of any interactions and effects of humidity, temperature, agent type, agent concentration, dynamic profile, and chemical interferent type and concentration
- Assessment of potential for false-positive and false-negative situations
- Probability of detection and identification of each agent (class and type) or simulant as a function of agent or simulant concentration
- Probability of a false alarm and estimate of mean time between false alarms

The U.S. government supports research of chemical sensors mainly through the Department of Defense, National Science Foundation, and the Department of Energy.

Sensor development is also heavily supported by private industry. Most of the current technological advances have had little real impact on improving emergency preparedness. Emergency preparedness requires that sensors be reasonably inexpensive so that they can be widely deployable and networked. More importantly, sensors must be reliable with good sensitivity and specificity toward detection of the targeted substances. Development of sensors that can detect and identify the release of toxic materials must continue. Effective responses to the specific agent involved in a chemical attack can only be achieved through the right choice of sensors for the job. Proper detection equipment is crucial for effective consequential management including orderly evacuation to minimize casualties. Therefore, a program with sustained funding to focus and coordinate research and development on sensors and sensor networks together with an emphasis on fielding the system is needed.

CHAPTER 2

Chemical Warfare Agents and Toxic Industrial Chemicals

CWAs are chemicals that are intended for use in warfare or terrorist activities with the intention to kill, seriously injure, or incapacitate people through physiological effects upon exposure. These chemicals produce disabling effects that severely reduce the ability to fight. When CWAs enter or contact the human body, they attack various organs and prevent them from functioning normally. CWA classifications, defined in Chapter 1, include nerve, blister, choking, blood, incapacitating, and vomiting agents. Nerve, blister, and blood agents are considered the most noxious or dangerous because a very small dose may cause death.

TICs are chemicals toxic to humans that are widely used in manufacturing or primary material processing (mining and refining). To qualify as a TIC, the chemical must have an LCt_{50} value of less than 100,000 mg-min/m^3, and be produced at a level of more than 30 tons per year at a single production facility. LCt_{50} means 50% of the exposed population would die when that dosage (concentration multiplied by exposure time) is ingested by an average-sized individual. A number of chemicals (more than 90), including many household-cleaning agents (e.g., ammonia and chloride in bleach), are considered TICs. Several of the TICs, such as hydrogen cyanide, cyanogen chloride, and phosgene, were once classified as chemical agents by the military, and were used in CW attacks during the world wars. TICs are ranked as being high-, medium-, or low-hazard TICs based on the combination of their toxicity level, amount being produced, and relative volatility.

Improper handling and accidents or intentional use on civilian targets by terrorists can lead to disasters. (Toxic chemical leaks and spills around the world are reported on the Environmental Protection Agency [EPA] website at http://yosemite.epa.gov/oswer/ceppoweb.nsf/content/ap-chsa.htm.)

One example of such disasters, based on the EPA Office of Solid Waste and Emergency Response, July 1993 (EPA-560-K-93-001 Series 8, No. 3), is the release of 30,000 to 50,000 pounds of hydrogen fluoride (HF) and isobutane at a Texas petroleum refinery due to a construction accident at an alkylation unit. The chemical vapors migrated to an adjacent residential area. Eighty-five square blocks and approximately 4000 residents were evacuated. Although no fatalities were reported, more than 1000 residents sought treatment at three local hospitals.

2.1 CHEMICAL, PHYSICAL, AND TOXICOLOGICAL PROPERTIES

Toxicity of chemicals can be expressed in terms of IDLH (immediate danger to life and health) level, LCt_{50}, REL (recommended exposure limit) and PEL (permissible exposure limit), and AEL (acceptable exposure limit), among others. Physical and chemical properties have direct influence on the behavior of a compound. For example, a vapor or liquid with high vapor pressure would create a respiratory inhalation hazard as the primary threat. Solids or low-volatility liquids may not produce a sufficiently concentrated vapor to act as an inhalation threat. Their primary threat would be through direct contact. Specific hazards will dictate respective detection and evaluation methodologies.

2.1.1 Molecular Formula

A molecular formula shows the molecular composition of a substance. It provides the actual number of atoms of each element in a molecule of the compound. The formula of sarin, for example, is $C_4H_{10}FO_2P$. This formula shows that one sarin molecule contains 4 carbon atoms, 10 hydrogen atoms, 1 fluorine atom, 2 oxygen atoms, and 1 phosphorus atom. The formulas for CWAs and some of the most hazardous TICs are included in this chapter's tables.

2.1.2 Molecular Structure Formula

A molecular structure formula differs from a molecular formula in that it indicates the relative connectivity or arrangement of the atoms in a molecule. For example, the molecular structure of sarin with the molecular formula of $C_4H_{10}FO_2P$ is shown Figure 2.1:

The structural formula shows how atoms are grouped and bonded to form the sarin molecule. This formula provides clues for predicting the substance's behavior.

Organic compounds with the same molecular formula, but different structures, are called isomers. Isomers' properties may be very different. For example, ethanol and dimethyl ether have the same molecular formula, C_2H_6O. Ethanol is a colorless

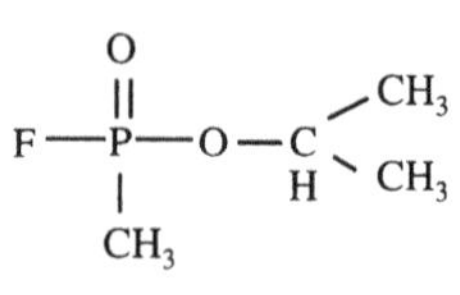

Figure 2.1 Structure of GB.

$$CH_3\text{—}CH_2\text{—}OH$$

a. Ethanol

$$CH_3\text{—}O\text{—}CH_3$$

b. Dimethyl ether

Figure 2.2 Structures of (a) ethanol and (b) dimethyl ether.

liquid; its structural formula is shown in Figure 2.2a. Dimethyl ether is a colorless gas; its structural formula is shown in Figure 2.2b.

Compounds with different formulas may have similar physical/chemical behaviors when their molecules are in the same function group. For instance, both hydrogen cyanide (H-CN) and cyanogen chloride (CN-Cl) contain the "cyanide (CN)" function group. Therefore, they act similarly to produce similar effects upon contacting blood cells.

2.1.3 Molecular Weight

Molecular weight (MW) is the weight of a molecule expressed in atomic mass units (amu), and calculated from the molecular formula of a substance. It is the sum of the atomic weights of the atoms that comprise the molecule. High-molecular-weight substances tend to be solids. Low-molecular-weight substances tend to be gases. MW also indicates the relative persistency of a compound. Heavier molecules usually have lower rate of evaporation and, thus, greater persistency.

2.1.4 Relative Vapor Density

Relative vapor density is the ratio between the density of a gas or vapor (D_g) and the density of air (D_a) at the same temperature and pressure conditions:

$$VD = \frac{D_g}{D_a} \tag{2.1}$$

Vapor density of a chemical is directly proportional to its MW. The average MW of air is 29 amu. At 25°C and 1 atmospheric pressure, average air density is 1.19 kg/m^3. The vapor density of air is referred to as being 1 for the calculation of relative vapor density of other vapors. Any chemical with MW >29 amu would have a vapor density >1 under the same temperature and pressure conditions, according to the Ideal Gas Law. Hence, the vapor is heavier than air and it will tend to stay in or flow to low places. Vapor of chemicals with MW <29 amu have a vapor density of <1. These vapors would tend to migrate to higher places and will be easily diluted by air currents. The vapor densities of all CWAs and TICs, except for a few chemicals such as hydrogen cyanide, carbon monoxide, diborane, and ammonia, are >1.

2.1.5 Vapor Pressure

Vapor pressure (VP) is the pressure generated by the vapor of a substance that is at equilibrium with its liquid and/or solid state in an enclosed space. In other words, if there is only one chemical in a closed container, and the evaporation/condensation between the vapor and the liquid/solid has been established, the pressure on the wall of the container will be generated by that chemical only. Vapor pressure determines the saturated concentration of a given vapor in air (volatility) and the chemical's persistency. A chemical with VP lower than the ambient air pressure evaporates gradually at a certain rate (such as water evaporation at room temperature). Lower VP indicates lower volatility, and hence, higher persistency. A chemical is gaseous when its VP is equal or greater than the atmospheric pressure (e.g., oxygen and nitrogen in the air, or water vapor when water is heated to its boiling point). Nerve and blister agents have VPs of <1 atmospheric pressure (760 mmHg), while most of the high-hazard TICs have VPs of >1 atmospheric pressure.

2.1.6 Volatility

When the vapor of a chemical is at equilibrium with its solid or liquid phase, the total weight of the chemical vapor found in terms of milligrams (mg) in a cubic meter (m^3) volume of air at a specified temperature represents its volatility. Its units are normally expressed as mg/m^3 at a specific temperature. Volatility (V) represents the maximum amount of a chemical that a cubic meter air can hold at the specified temperature. For example, the volatility of water at 20°C is 23,010 mg/m^3. This means that at that temperature, 1 cubic meter air contains 23.01 g of water when the air is saturated.

In an open space, the chemical vapor will continually be diluted by air and diffuse away. The vapor–solid or vapor–liquid equilibrium can never be reached. Thus, concentration of a compound is at its maximum near the interface. V is directly related to temperature. At a higher temperature, a compound will have relatively higher volatility. This is best illustrated by the different amounts of water that air can contain at different temperatures. During the summer, humidity (water content in the air) can be much higher than during the winter. V is one of the most important physical properties of an agent or TIC that one should know when dealing with detection device development, vapor generation, and equipment evaluations.

V plays an important role in determining deployment and detection effectiveness. The V of all CWAs and some TICs is higher or much higher than their respective IDLH concentration levels. This suggests that the chemicals can easily reach dangerous vapor concentrations in the surrounding air.

For chemicals of similar toxicity, higher-volatility compounds are considered more dangerous than lower-volatility ones. That is because highly volatile substances enter the air more easily and will produce higher vapor concentrations. Methods designed to detect chemical vapors must treat volatility as a first consideration. Methods required for detection of chemicals with very low volatility as opposed to those with higher volatility may be different. Similarly, a thorough understanding of the compound's volatility is a must for proper vapor generation, sample collection, safe handling, and effective protection or deployment of these types of chemicals. Its importance cannot be over-emphasized.

V is inversely related to the persistency of a chemical. A chemical with high persistency, and thus lower volatility, will be found on the surface of an object long after the initial encounter. Residues of persistent blister agents and VX enter the human body through absorption on contact with unprotected skin. Their persistency enhances their effectiveness. The vapor concentration of highly persistent chemicals in the air is likely to be quite low. Such low concentration availability makes detection of these chemicals in the atmosphere more difficult. A detection instrument must be able to identify the targeted compound in highly diluted states; that is, the device must segregate a minute amount of material in an ocean of potentially interfering substances. Therefore, surface-sampling techniques should be used to supplement vapor detection techniques when detection of such chemicals is desired.

2.1.7 Concentration

Concentration refers to the amount of a substance contained in a volume of liquid or vapor. Since detection of toxic chemical vapors is our primary objective, our attention will be focused on those dealing with vapor concentrations. Vapor concentration is usually expressed in terms of number of milligrams per cubic meter (mg/m^3). Some practitioners may use different terms such as microgram per liter ($\mu g/l$), parts per million (ppm) or parts per billion (ppb), or percentage (for high concentrations). When concentration is expressed using parts per million or billion, temperature and pressure must be specified. Milligrams per cubic meter or micrograms per liter denotes that the value is the actual vapor concentration that contains the same amount of the substance in a fixed volume no matter what the temperature and atmospheric pressure are, while the actual amount of an agent in vapor varies with temperature and pressure when the concentration is expressed as parts per million or billion. More details on concentration units and conversion among them are provided in Chapter 4.

2.2 TOXIC PROPERTIES

Toxicity together with the volatility of a compound dictates the effectiveness of a CWA or TIC when deployed. Common terms used to gauge the relative toxicity of a CWA or a TIC are IDLH level, LCt_{50}, $ICt_{50,}$ TWA (time-weight average), and AEL.

2.2.1 IDLH Level

IDLH level is the concentration of a chemical in the air that would cause immediate or delayed permanent adverse heath effects after 30 min of unprotected exposure. At IDLH level, an exposed person may have 30 min to escape the contaminated environment without loss of life or irreversible health effects. The IDLH level was established by the National Institute for Occupational Safety and Health (NIOSH) as respirator selection criteria in the mid-1970s.

The purpose for establishing the IDLH level is to determine a concentration that permits sufficient time for a person to escape safely or to seek protection without incurring serious injury or irreversible health effects. In determining IDLH levels, the ability of a person to escape or seek protection was considered together with severe eye or respiratory irritation and other deleterious effects (e.g., disorientation or lack of coordination) that could prevent a timely escape. The IDLH level is the concentration at which the dosage received during a 30-min exposure without physical protection would cause minimal adverse health effects. However, the 30-min period is *not* meant to imply that people should stay in that contaminated environment any longer than necessary. In fact, every effort should be made to evacuate the locale immediately.

IDLH levels vary widely among diverse compounds. A lower value means that the compound is more dangerous and has greater toxicological effects. The volatility of all CWAs and most TICs is, unfortunately, much higher than respective determined IDLH level. This means that the vapor concentration in the air can easily reach the dangerous level.

The establishment of IDLH levels does not mean that exposure to lower levels of CWAs or most TICs is safe. Exposure to chemicals at concentrations much lower than the IDLH level for a long period will also be dangerous or even fatal. Although exposures to lower concentrations for a short period may not cause immediately observable effects or symptoms, toxicity effects may be cumulative. While some toxic substances (such as HCN) can be gradually detoxified naturally by human organs, many of them cause irreversible effects. The latter type of compound is commonly called "refractory." Soman (or GD) and other refractory compounds must be handled with special care to avoid any exposure.

2.2.2 LCt_{50}

LCt_{50} represents the lethal (L) concentration (C) that will cause death to 50% of the exposed population for a period of exposure time (t). It is a function of vapor concentration (C = mg/m^3) and exposure duration (t in minutes).

LCt_{50} values are established for both inhalation and percutaneous exposures. Inhalation dosage is exposure through the respiratory tracts. Percutaneous exposures occur through skin contact. For most chemicals of interest here, the inhalation LCt_{50} dosage is lower than that of the percutaneous exposure because the body absorbs the chemical vapor much faster and more effectively through the respiratory tract than through skin permeation. Thus, it is most important to protect the airway immediately to minimize inhalation dosage. Once the airway has been protected, it is necessary then to eliminate contaminated clothing and remove residue contamination as soon as possible. In general, absorption through the skin is somewhat limited to exposed skin initially, and then later the chemical penetrates clothing.

Dosage is a function of concentration and exposure time. For example, if 100 people exposed to air containing 100 mg/m^3 of a chemical that has an LCt_{50} value

Table 2.1 Concentrations and Exposure Time to Reach LCt_{50} of Sarin via Inhalation

Concentration of Sarin in Air (mg/m^3)	Exposure Time for Person at Rest (min)[a]	Exposure Time for Mildly Active Person (min)[b]
200	0.5	0.35
100	1	0.7
50	2	1.4
5	40	14

[a] LCt_{50} is 100 mg-min/m^3 for person at rest.
[b] LCt_{50} is 70 mg-min/m^3 for mildly active person.

of 100 mg-min/m^3 without protection, about 50 people would be killed with 1 min of exposure. Fifty people will be killed with 4 min of exposure when the concentration is 25 mg/m^3. Table 2.1 shows exposure times for sarin to reach LCt_{50} at various concentrations.

A few factors may affect LCt_{50}. Respiration rate is one of them. The respiration rate is slower when an individual is resting than when active; a smaller dosage will be breathed in during the same period of time and at the same concentration level when a person is at rest. A resting individual will also have slower blood circulation due to a lower pulse rate. Lower breathing and pulse rates mean lower dosage ingested over a given time period and at a given concentration. Therefore, the LCt_{50} value is generally higher for a person at rest (e.g., 100 mg-min/m^3 for GB) than for an active person (e.g., 70 mg-min/m^3 for GB).

Similar to LCt_{50}, the concentration that incapacitates 50% of the exposed population multiplied by the exposure time is defined as ICt_{50}. The ICt_{50} value is lower than the LCt_{50} because it is the dosage that will cause incapacitation of an individual to perform normal functions effectively rather than to cause death. For example, if exposure to a chemical compound causes coughing, vomiting, hallucinating, or any uncontrolled bodily functions, it will have diminished the effectiveness of an individual to perform normal functions. If the individual were a soldier, he or she would be spending time dealing with symptoms instead of fighting.

Given that people ingest lower doses when at rest, one of the most important keys to surviving a chemical attack is to remain calm.

2.2.3 TWA

The concentration of toxic chemicals may vary during the period of exposure. TWA reflects the exposure level over a period time. For example, assume that a person is exposed to a chemical at a concentration of 0.1 ppm for 6 hr and at a concentration of 0.2 ppm for 2 hr. The TWA over the 8-hr period is calculated as

$$\text{TWA} = (0.1 \text{ ppm} \times 6 \text{ hr} + 0.3 \text{ ppm} \times 2 \text{ hr}) / 8 \text{ hr} = 0.15 \text{ ppm} \quad (2.2)$$

Three common terms are based on TWA. REL, or recommended exposure limit, is maximum TWA concentrations for up to a 10-hr workday during a 40-hr workweek. PEL, or permissible exposure limit, is the TWA concentration that must not be exceeded during any 8-hr workday of a 40-hr workweek. STEL, or short-term exposure limit, is designed for a 15-min TWA exposure that should not be exceeded at any time during a workday.

2.2.4 Other Terms

Other terms that may be used include LC_{50} (lethal concentration for 50% of population), IC_{50} (incapacitation concentration for 50% of population), and LD_{50} (lethal dosage for 50% of population). These terms will be explained when they are used.

2.3 COMMON CWAs

Included here are nerve agents, blister agents, choking agents, and blood agents, among others. The important agents in each group, their common names, symbols, and symptoms upon exposure are summarized in Table 2.2. Each type of agent is briefly reviewed below.

2.3.1 Nerve Agents

Nerve agents, including the G- and the V-agents, attack the nervous system by disrupting the normal transmission of nerve impulses. Since muscle functions depend on nerve signals, interrupting signal transmission produces neurologic disorders, followed by paralysis, cardiovascular, or respiration failure. Lungs and eyes could

Table 2.2 Common Agents and Symptoms After Exposure

Group	Common Agents	Common Symptoms After Exposure
Nerve agents	GA (tabun), GB (sarin), GD (soman), GF (cyclosarin), and VX	Runny nose, tightness of chest, dimness of vision and myosis, breathing difficulty, drooling and excessive sweating, nausea/vomiting, involuntary defecation and urination, twitching, jerking and staggering, headache, confusion, drowsiness, convulsions, coma, and eventually, death. Death is caused by respiratory arrest, as all respiratory muscles contract
Blister agents	Distilled mustard (HD), nitrogen mustards (HN-1, HN-2, HN-3), lewisite (L)	Reddening of skin that develops into inflammation and/or blister
Choking agents	Phosgene (CG), diphosgene (DP)	Sneezing, pain in chest, painful swallowing, choking, breathing difficulty, and tightness in chest
Blood agents	Hydrogen cyanide (AC), cyanogen chloride (CK)	Convulsions, rapid deep breathing, and bradycardia followed by shallow breathing and eventual cessation of breathing. CK has choking and strong irritation effects

rapidly absorb nerve agent vapor, resulting in deadly effect in <1 min. Reaching lethal dosage occurs rapidly after exposure to liquid nerve agents as well.

Nerve agents cause an increase in acetylcholine throughout the body. Acetylcholine interferes with the vital enzyme, cholinesterase. Thus, nerve agents are also known as cholinesterase inhibitors. Acetylcholine and cholinesterase play a vital role in controlling nerve signal transmission to skeletal muscles, autonomic ganglia, and many central nervous system structures. Symptoms include difficulties in breathing, drooling and excessive sweating, nausea, vomiting, and cramps, among others. Nerve agents are extremely toxic. Death can occur within minutes when a sufficient dose enters the body through the respiratory route. Exposure symptoms appear more slowly from skin exposures. Although a lethal dose may be reached in 1 to 2 min of exposure, death may be delayed for 1 to 2 hr.

G-agents such as tabun (GA), sarin (GB), soman (GD), and cyclosarin (GF) are fluorine- or cyanide-containing organophosphates, and V-agents (VX and Vx) are sulfur-containing organophosphorus compounds. They are colorless liquids at 20°C, with or without odors. The MWs of the nerve agents are >29 amu; therefore, their vapors are heavier than air (relative vapor density >1). Among the nerve agents, GB has the highest VP, and VX the lowest. VP of the nerve agents decreases in the following order:

$$GB > GD > GF > GA >> VX$$

The toxicity of VX is the highest among the nerve agents with an IDLH level of 0.002 ppm. It is about 10 times more toxic than GB and GA. GD is the most toxic agent among the G-agents. The IDLH level for nerve agents decreases in the following order:

$$VX > GD > GA, GB, GF$$

Table 2.3 provides toxicity data for nerve agents, including IDLH, LCt_{50}, and ICt_{50}. IDLH levels are listed as ppm. However, these indicators can be calculated as milligrams per cubic meter, as discussed in Chapter 3.

2.3.1.1 Tabun

The first nerve agent, tabun, also known as GA, was discovered by German scientists who had developed the fluorine-containing compound as a pesticide. The effect of nerve agents on human beings was first noticed when workers were observed to experience pupil contraction and shortness of breath when exposed to tabun vapor.

GA, with the structure formula shown in Figure 2.3 — chemically known as ethyl N,N-dimethylphosphoroamidocyanidate — is a colorless to brownish liquid. Its vapor is also colorless. GA may have a faintly fruity odor when not pure. It is about 30 times as toxic as phosgene, which was used in WWI. Like other G-agents, it enters the body primarily through the respiratory tract. GA is approximately 20 times more persistent than GB, but not as stable in storage. At 20°C, its volatility is 328 mg/m^3, which increases to 858 mg/m^3 when temperature changes to 30°C.

Table 2.3 Properties of Common Nerve Agents

Agent	Molecular Formula	CAS #	MW (amu)	Vapor Pressure (mmHg at 20°C)	Volatility (mg/m^3 at 20°C)	Vapor Density	Odor	IDLH (ppm)	LCt_{50} ($mg\text{-}min/m^3$)[a]	ICt_{50} ($mg\text{-}min/m^3$)[a]
GA	$C_5H_{11}N_2O_2P$	77-81-6	162.13	0.037	328	5.63	Faintly fruity	0.03	400 (Resting)	300 (Resting)
GB	$C_4H_{10}FO_2P$	107-44-8	140.10	2.10	16,091	4.86	None	0.03	100 (Resting)	75 (Resting)
GD	$C_7H_{16}FO_2P$	96-64-0	182.12	0.40 (25°C)	3,900 (25°C)	6.33	Camphor	0.008	70 (Mild activity)	In GA and GB range
GF	$C_7H_{14}FO_2P$	329-99-7	180.2	0.044	438	6.2	None	0.03	—	—
VX	$C_{11}H_{26}NO_2PS$	50782-69-9	267.38	0.0007	10.5 (25°C)	9.2	None	0.002	100	50 (Resting)

[a] For respiration exposure.

CAS #, Chemical Abstracts Service registry number; IDLH, immediate danger to life and health; MW, molecular weight.

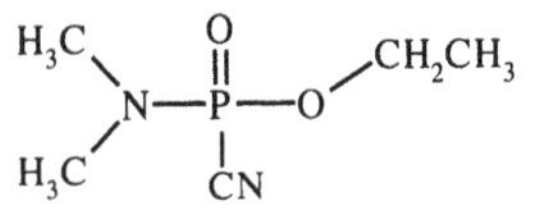

GA: Ethyl N, N-dimethylphosphoroamidocyanidate

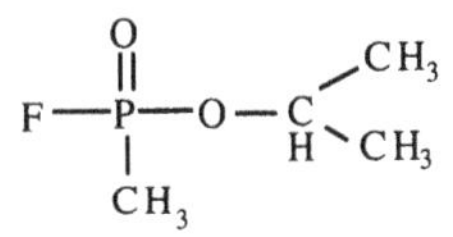

GB: Isopropyl methylphosphonofluoridate

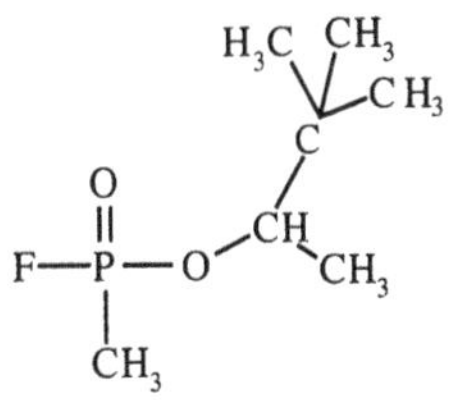

GD: Pinacolyl methyl phosphonofluoridate

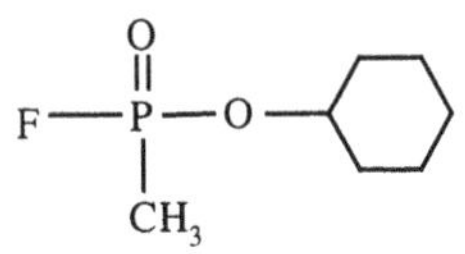

GF: O-Cyclohexyl-methylfluorophosphonate

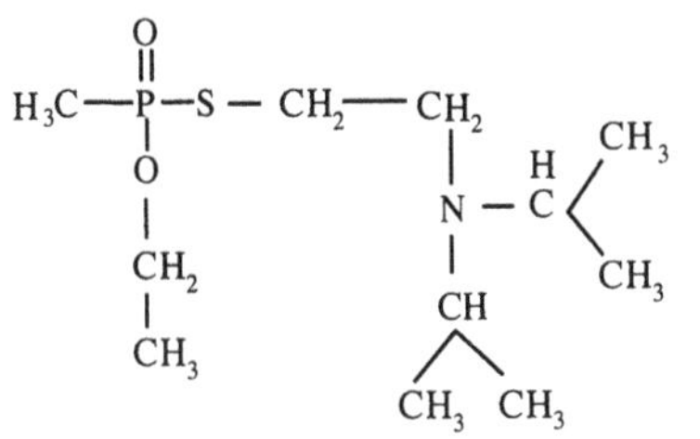

VX: O-ethyl-S-(2-iisopropylaminoethyl)methyl phosphonothiolate

Figure 2.3 Structure and chemical names of nerve agents.

2.3.1.2 Sarin

The second nerve agent, sarin, also known as GB, was discovered in 1938. Its structure formula is given as Figure 2.2 together with its chemical name. GB is a potent organophosphate with an IDLH level similar to GA at 0.03 ppm. It is a colorless liquid at ambient temperatures with almost no odor in its pure state. Its volatility at 20°C is 16,091 mg/m^3, which is much higher than that of GA (328 mg/m^3). GB molecules contain a fluorine atom (–F) instead of a –CN group, as in GA molecules.

2.3.1.3 Soman and Cyclosarin

Subsequent to the discoveries of GA and GB, soman, also known as GD, was discovered in 1944. Details of the discovery were uncovered by the Soviets, who subsequently produced and stockpiled soman. GD molecules may penetrate the central nervous system within minutes, which makes GD the most toxic of the G-agents. GD is also a colorless liquid at ambient temperature that evaporates to colorless vapor with a fruity odor. Its IDLH level is 0.008 ppm, which is much lower than other G-agents. Soman volatility of 3900 mg/m^3 at 25°C is also much higher than that of GA. Since it permanently binds to receptors in the body, it is considered a refractory agent. Similar to GB, the GD molecule also contains a fluorine atom (–F).

Cyclosarin, or GF, is another G-agent that contains a fluorine atom. Its IDLH level is 0.03 ppm, similar to GA. GF volatility (438 mg/m^3) is higher than that of GA, but much lower than that of GB and GD.

The abovementioned chemicals are G-agents whose molecules contain phosphorous and either a fluorine (–F) or cyanide (–CN) function group. Molecules of the other group of nerve agents, V-agents, contain both sulfur and phosphorous atoms.

2.3.1.4 VX and Vx

VX, or S-{2-[bis(1-methyl)amino]ethyl}-O-ethyl methylphosphonothiolate, is one of a group of organophosphate esters derived from substituted 2-aminoethanethiols. Soviet scientists developed a slightly different compound — S-[(diethylamino)ethyl]-O-ethyl isobutylphosphonothiolate — with the same molecular formula as VX, known as V-gas or Vx. These compounds were discovered in the 1950s. They are more toxic than the G-class compounds. They have much lower volatility, and thus are more persistent but less effective in causing respiratory exposure casualties than G-agents. Except for direct exposure to an active aerosol shortly after dissemination, the primary hazard from this class of nerve agents is through contact with contaminated surfaces after aerosols have settled.

2.3.2 Blister Agents

Blister agents blister the skin and produce chemical burns to any exposed part of the body. Eyes, mucous membranes, and lungs are especially sensitive. Unlike nerve agents, which can kill exposed people rapidly, blister agents are primarily intended to injure rather than kill. Injured people require attention by responders, and thus tie up more resources than dead people. Large-scale severe injuries or slow deaths are strategically more effective in defeating the enemy soldiers' efficiency.

Blister agents include arsenicals and mustards. Lewisite (L) is the most important arsenical. Mustards include sulfur mustard (HD) and nitrogen mustards (HN-1, HN-2, and HN-3); of these, HD and HN-3 are the most common.

Figure 2.4 provides the structures and chemical names of selected blister agents and Table 2.4 lists the chemical and physical properties of blister agents. All blister agents are liquids at 20°C, with characteristic odors. The mustards are colorless to amber-colored, and lewisite ranges from a clear to brown liquid. HD, HN-3, and L are lower in VP than HN-1 and HN-2. The vapor density of each of the blister agents is heavier than air.

Blister agents are less toxic than nerve agents. There are no IDLH levels established for them due to potential carcinogenic properties of these compounds. LCt_{50} and ICt_{50} levels are also listed in Table 2.4.

Mustards may penetrate exposed skin in 1 or 2 min, but symptoms may not be noticeable for 2 hr and up to 48 hr later. Unlike mustards, lewisite causes immediate

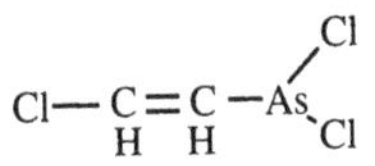

L: Dichloro-(2-chlorovinyl)arsine

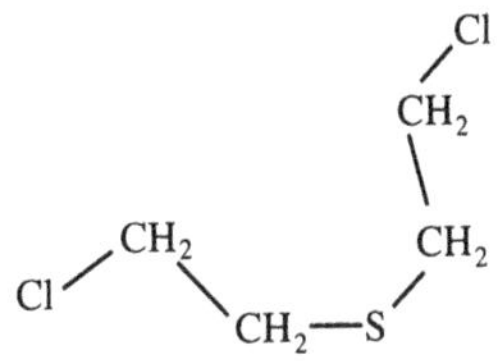

HD: Bis-(2-chloroethyl) sulfide

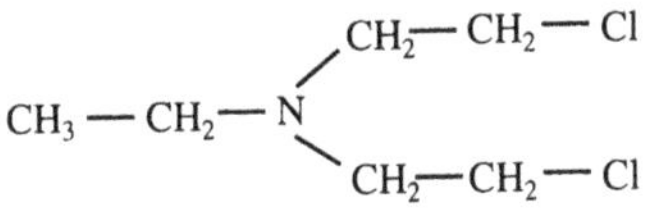

HN-1: 2, 2-Dichlorotriethylamine

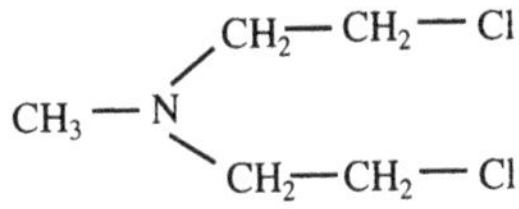

HN-2: 2, 2' -Dichloro-N-methyldiethylamine

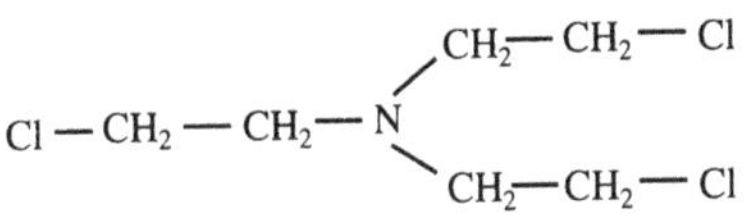

HN-3: 2, 2', 2" - Trichlorotriethylamine

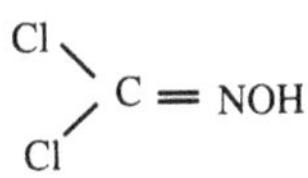

CX: Dichloroformoxime

Figure 2.4 Structure and chemical names of blister agents.

pain upon contact. The damage caused by blister agents may take up to several months to heal. Severity of the injury caused by blister agents is directly related to concentration of the agent, route of entry, and duration of exposure.

2.3.2.1 Arsenicals

Arsenical compounds contain arsenic as a central atom. These blistering compounds are derived from replacing the hydrogen atoms of the blood agent arsine (AsH_3) with various organic radicals, chloride, or cyanide, including lewisite (L, $C_2H_2AsCl_3$), mustard lewisite mixture (HL), phenyl dichloroarsine (PD, $C_6H_5AsCl_2$), ethyldichloroarsine (ED, $C_2H_5AsCl_2$), and methyldichloroarsine (MD, CH_3AsCl_2).

2.3.2.2 Levinstein Mustard and Distilled Mustard

Levinstein mustard (H) is made using the Levinstein process. Approximately 30% of H is sulfur, which explains its pronounced odor. When H is purified, the resulting compound is designated as distilled mustard (HD). HD has less of a

Table 2.4 Properties of Common Blister Agents

Agent	Molecular Formula	CAS #	Molecular Weight	Vapor Pressure (mmHg)[a]	Volatility (mg/m³)[a]	Vapor Density	Odor	LCt_{50} (mg-min/m³) Respiration	LCt_{50} (mg-min/m³) Percutaneous	ICt_{50} (mg-min/m³) Respiration	ICt_{50} (mg-min/m³) Percutaneous
HD	$C_4H_8Cl_2S$	505-60-2	159.08	0.072	610	5.4	Garlic	1500	10,000	150	2000
HN-1	$C_6H_{13}Cl_2N$	538-07-8	170.08	0.24	1520	5.9	Faintly fishy or musty	1500	20,000	—	9000
HN-2	$C_5H_{11}Cl_2N$	51-75-2	156.07	0.29	3580 (25°C)	5.4	Fruity odor in high concent-ration	3000	—	—	Between HN-1 and HN-3
HN-3	$C_6H_{12}Cl_3N$	555-77-1	204.54	0.0109 (25°C)	121 (25°C)	7.1	None	1500	10,000	—	2500
L	$C_2H_2AsCl_2$	541-25-3	207.35	0.087	4480	7.1	Geranium	1400	100,000	—	>1500

[a] At 20°C.

CAS #, Chemical Abstracts Service registry number.

garlic-like odor, is more stable, and has slightly greater blistering power than H. Moist skin absorbs mustard more readily than dry skin. Thus, mustard used in hot and humid weather will result in a higher casualty rate. Eyes are extremely susceptible to mustard exposures. Very low levels of exposure to mustard gas will cause "red-eye" conjunctivitis. High-level exposure results in strong lung irritation and overall disabling effects. The detoxification rate is very slow. Due to sensitization, repeated exposures are cumulative. Mustard is also found to be a carcinogenic compound. Consequently, there is no exposure concentration level value considered as safe for human health.

2.3.2.3 Nitrogen Mustards

Nitrogen mustards (HN) are compounds derived from ammonia by replacing hydrogen atoms with various organic groups. The central atom is nitrogen. Included in this group are 2,2-Dichlorotriethlyamine (HN-1), 2,2'-Dichloro-N-methyldiethylamine (HN-2), and 2,2',2"-trichlorotriethylamine (HN-3). Physiologic effects of nitrogen mustards are similar to those of HD. HN compounds are also useful in treating certain cancers.

HN-1 is a colorless liquid with a faint, fishy or musty odor. It is used as a delayed-action agent; damage to skin begins to appear 12 or more hr after exposure. However, HD may act more quickly on the eyes than HD. The body does not detoxify HN-1, and it is a refractory agent.

HN-2 may appear as a dark liquid with a fruity odor at high concentrations and a soft soap odor at low concentrations. Its toxicity is somewhat higher than HN-1. HN-2 has the highest blistering power of all nitrogen mustards when in vapor form. Skin effects for this compound are delayed, and it is highly unstable.

HN-3, a liquid with no odor in its pure form, is the most common nitrogen mustard and its blistering power is similar to that of HD. It is used as a delayed-action casualty agent — most symptoms may be delayed for 4 to 6 hr. The body does not detoxify HN-3; therefore, it accumulates in the body. Unlike HN-2, HN-3 is stable in storage.

2.3.2.4 Mustard-T Mixture

Mustard-T (HT) is a mixture of 60% HD with 40% T (Bis-[2(2-chloroethylthio)ethyl] ether, $(ClCH_2CH_2SCH_2CH_2)_2O$). HT contains sulfur and chlorine similar to the HD structure. It is a yellowish liquid having an odor similar to that of HD. HT possesses a stronger blistering effect than HD. It is more stable and has a lower freezing point than HD.

2.3.2.5 Mustard-Lewisite Mixture

Mustard-lewisite mixture (HL) is a variable mixture of HD and L. The mixture ratio varies according to weather conditions during an operation. L is added to the HD to increase volatility and to lower the freezing point of HD to permit cold-weather application.

2.3.2.6 Phosgene Oxime

Phosgene oxime (CX) may appear as a colorless, low-melting-point crystalline solid or as a liquid. It has a high VP and decomposes slowly at normal temperatures. CX is readily soluble in water, and possesses a disagreeable and penetrating odor. Upon contact with the skin, it produces immediate pain resembling a bee sting. It causes violent irritation to mucous membranes of the eyes and nose. Rapid reaction with the skin makes it difficult for effective decontamination. Flushing with large amounts of water immediately upon exposure proves to be the most effective way to quickly remove excess agent from the skin.

2.3.3 Blood Agents

Blood agents, which enter the body primarily through the respiratory tract, prevent the normal utilization of oxygen by blood cells. Included are hydrogen cyanide (AC), cyanogen chloride (CK), and arsine (SA). Figure 2.5 shows blood agent structures, and Table 2.5 shows properties.

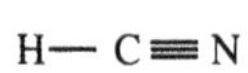

AC: Hydrogen cyanide

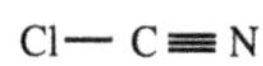

CK: Cyanogen choride

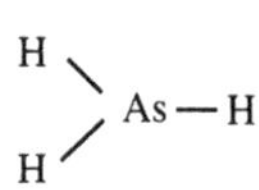

SA: Arsine

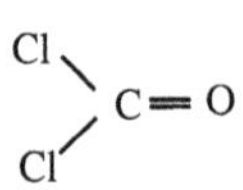

CG: Carbonyl chloride

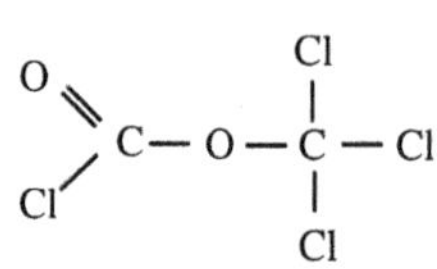

DP: Trichloromethyl chloroformate

Figure 2.5 Structure and chemical names of blood and choking agents.

Table 2.5 Properties of Common Blood and Choking Agents

Agent	Molecular Formula	CAS #	Molecular Weight	Vapor Pressure (mmHg)[a]	Volatility (mg/m³)[a]	Vapor Density	Odor	LCt_{50} (mg-min/m³)	ICt_{50} (mg-min/m³)
AC	HCN	74-90-8	27.03	612	1,080,000 (25°C)	0.99	Similar to bitter almonds	2000 to 4500[b]	Varies with concentration
CK	CNCl	506-77-4	61.48	1010	Gas	2.2	Pungent, biting	11,000	7000
SA	AsH_3	7784-42-1	77.93	11,100	30,900,000 (0°C)	2.69	Mild, garlic like	5000	2500
CG	$COCl_2$	75-44-5	98.92	1173	4,300,000 (7.6°C)	3.4	Similar to new-mown hay, grass, or green corn	3200	1600
DP	$C_2Cl_4O_2$	503-38-8	197.85	4.2	45,000	6.8	New-mown hay, grain, or green corn	3000 (Resting)	1600 (Resting)

[a] At 20°C.

[b] LCt_{50} and ICt_{50} vary with concentration because of the high detoxification rate at which the body decomposes HCN to nontoxic compounds. At a concentration of 200 mg/m³, the LCt_{50} is 2000. At 150 mg/m³, the LCt_{50} is 4500.

CAS #, Chemical Abstracts Service registry number.

2.3.3.1 *Hydrogen Cyanide and Cyanogen Chloride*

AC and CK are colorless liquids with distinctive odors. The vapor density of AC is slightly less than that of air while that of CK is heavier. AC is more toxic than CK due to higher cyanide content, as indicated by respective IDLH levels and LCt_{50} values.

AC inhibits the cytochrome oxidase enzyme reaction, which directs oxygen utilization in the bloodstream. Upon exposure, the breathing rate increases, which in turn leads to inhalation of a larger dose. At high concentrations, exposed individuals could be killed in less than 15 sec. Because the body can detoxify AC quickly, its LCt_{50} varies according to exposure concentration. The LCt_{50} of 2000 mg-min/m^3 in Table 2.5 represents the concentration of AC at 200 mg/m^3. CK affects the oxygen utilization function similar to AC. In contrast to AC, CK causes breathing to slow down because of its strong irritation and choking effects.

2.3.3.2 *Arsenic Trihydride*

AsH_3, also known as SA, damages the liver and kidneys. Light exposure causes headache and uneasiness. Higher-dose exposure causes chills, nausea, and vomiting. Severe exposure damages blood cells, leading to anemia and eventual death. SA has the highest VP and volatility and is the most rapidly dispersing chemical agent of the three common blood agents.

2.3.4 Choking Agents

Choking agents (Figure 2.5 and Table 2.5) target the respiratory tract and lungs. After entering the airway, the choking agent decomposes into hydrochloric acid and oxygen-free radicals, which irritate the respiratory tract, and cause membrane swelling. This swelling triggers the secretion of copious amounts of fluid, which in turn causes excessive coughing to clear the airway. Despite the coughing, the lungs may fill with fluid, and thus victims “choke,” while literally drowning in their own body fluid. Death occurs when dosage is high enough. The effects of choking agents may be immediate or delayed depending on exposure concentration.

These agents include phosgene (CG) and diphosgene (DP). In World War I, phosphene accounted for more than 80% of CWA fatalities. A gas at room temperature, phosgene boils at 7.6°C. Diphosgene, with much higher boiling point (127°C), is a colorless liquid at room temperature with new mown hay odor similar to that of phosgene. Both agents have similar IDLH and LCt_{50} values.

2.3.5 Other CWAs

Other chemical compounds are sometimes grouped according to their intended use, such as incapacitating agents and riot control agents.

Incapacitating agents produce physiologic or mental effects that may persist for several hours or days after an exposure. Such effects make people less effective fighters, but do not seriously endanger their lives, unless subjected to very high doses.

Compounds that cause vomiting are normally solids that are vaporized and condensed to form aerosols, including diphenylchloroarsine (DA), adamsite (DM), diphenylcyanoarsine (DC). Unless subjected to high-dose exposure, victims will recover in several hours.

Tear-producing compounds cause copious tears and irritation of the skin, and are widely used for training and riot control. The principal tearing compounds include chloroacetophenone (CN); chloroacetophenone in chloroform (CNC); chloroacetophenone, chloropicrin, and chloroform mixture (CNS); chloroacetophenone, benzene, and carbon tetrachloride mixture (CNB); bromobenzylcyanide (CA); and O-chlorobenzylidene malononitrile (CS).

The well-known "tear gas" consists of CS, CS1, and CS2. CS1 is comprised of CS mixed with silica aerogels and CS2 is the mixture of CS and silicone treated silica aerogels. CS in its pure form has a pepper-like odor, and produces immediate effects even in low concentrations. Maximum effects are reached in 20 to 60 sec and duration is 5 to 10 min after the individual is removed from the contaminated area.

2.4 TOXIC INDUSTRIAL COMPOUNDS

Toxic industrial compounds or chemicals (TICs) have received more attention in recent years mainly because of the fear of terrorist acts. Although they are much less deadly than conventional CWAs, terrorists can easily gain access to large quantities of TICs.

More than 90 chemicals are considered TICs, and they are grouped as high, medium, and low hazards based on the combination of toxicity, volatility, and amount being produced, stored, and transported annually. Each TIC is ranked according to a hazard index (HI). The HI is the product of four factors, numerically ranked between 1 and 5 according to the ranking scheme in Table 2.6.

$$HI = \text{toxicity} \times \text{state} \times \text{distribution} \times \text{producers} \qquad (2.3)$$

Table 2.6 Hazard Index Parameters

Toxicity (IDLH in ppm)	Index	State (Vapor Pressure in mmHg)	Index	Distribution[a]	Index	Number of Producers	Index
<1	5	Gas	5	5/6	5	>100	5
1 to 10	4	Liquid: >400	4	4	4	50 to 99	4
11 to 100	3	Liquid: 100 to <400	3	3	3	25 to 49	3
101 to 500	2	Liquid: 10 to <100	2	2	2	5 to 24	2
>500	1	Liquid: <10	1	1	1	<5	1

[a] Number of continents in which production occurs.

IDLH, immediate danger to life and health.

Source: U.S. Department of the Army. Appendix D of JCAD Performance Specification Revision B, 13 August 1999.

Chemicals with an HI value of ≥90 are listed in the "high" hazard group, those with an HI value between 36 and 89 are listed as "medium" hazards and those with an HI value <36 are considered "low" hazards.

2.4.1 High-Hazard TICs

High-hazard TICs (Table 2.7) are widely produced, stored, or transported chemicals with high toxicity that are easily vaporized. This group contains mainly inorganic chemicals except the two small molecular organic chemicals (formaldehyde and ethylene oxide). Some of the TICs in this group, such as chlorine, hydrogen cyanide, cyanogen chloride, and phosgene were once considered as CWAs by the military. Most of the chemicals in this group exist as gaseous compounds at room temperature as designated by "G" in the volatility column in Table 2.7. Volatility values range from 3.7 ppm to 800 ppm. Sulfuric acid has relatively low volatility but is an extremely strong acid; thus, it is considered a high-level contact hazard.

2.4.2 Medium-Hazard TICs

High-hazard TICs are highly toxic, produced in large quantities, and vaporize easily. TICs that may be ranked high in one or two of these characteristics, but low in the rest, are grouped as medium hazards. Table 2.8 lists properties of selected medium-hazard TICs.

The IDLH of the TICs in this group ranges from 1 ppm to 1200 ppm. Clearly, the volatility of each TIC in this group is higher than respective IDLH level. Vapor pressures are mostly lower than the compounds listed in the high-hazard group.

2.4.3 Low-Hazard TICs

Low-hazard TICs include compounds that have relatively low toxicity, VP, or volatility, and relatively small quantities are being produced, stored, or transported (Table 2.9). The threat from deployment of these TICs is considered much less likely than those listed as high or medium hazards. Nevertheless, their existence could still become threats to the public.

Table 2.7 Properties of High-Hazard TICs

Chemical	Formula	MW	Volatility	Odor	IDLH[a]	Effects
Ammonia	NH_3	17	G	Pungent, suffocating odor	300	Eyes, skin, respiratory system
Boron trichloride	BCl_3	117.2	G	Pungent	—	Corrosion in eyes, skin, and respiratory system
Boron trifluoride	BF_3	67.8	G	Pungent, suffocating odor	25	Eyes, skin, respiratory system, kidneys
Carbon disulfide	CS_2	76.1	390,000	Sweet ether-like odor	500	Central nervous system, peripheral nervous system, cardiovascular system, eyes, kidneys, liver, skin, reproductive system
Chlorine	Cl_2	70.91	19,369,000	Pungent, irritating odor	10	Eyes, skin, respiratory system
Diborane	B_2H_6	27.7	G	Repulsive, sweet odor	15	Respiratory system, central nervous system, liver, kidneys
Ethylene oxide	C_2H_4O	44.1	G	Ether-like odor	Ca [800]	Eyes, skin, respiratory system, liver, central nervous system, blood, kidneys, reproductive system
Fluorine	F_2	38.0	G	Pungent, irritating odor	25	Eyes, skin, respiratory system, liver, kidneys
Formaldehyde	HCHO	30	G	Pungent, suffocating odor	Ca [20]	Eyes, respiratory system
Hydrogen bromide	HBr	80.9	G	Sharp, irritating odor	30	Eyes, skin, respiratory system
Hydrogen chloride	HCl	36.5	G	Pungent, irritating odor	50	Eyes, skin, respiratory system
Hydrogen fluoride	HF	20.0	G	Strong, irritating odor	30	Eyes, skin, respiratory system, bones
Hydrogen sulfide	H_2S	34.1	G	Strong odor of rotten eggs	100	Eyes, respiratory system, central nervous system
Nitric acid (fuming)	HNO_3	63.0	63,000	Acrid, suffocating odor	25	Eyes, skin, respiratory system, teeth
Phosphorus trichloride	PCl_3	137.4	130,000	Odor like hydrochloric acid	25	Eyes, skin, respiratory system
Sulfur dioxide	SO_2	64.1	G	Characteristic irritating, pungent odor	100	Eyes, skin, respiratory system
Sulfuric acid	H_2SO_4	98.1	1.3	Odorless	3.7	Eyes, skin, respiratory system, teeth
Tungsten hexafluoride	WF_6	297.9	G	Odorless	NE[b]	Eyes, skin and mucous membranes

[a] The notation Ca appears in this column for all substances that the National Institute for Occupational Safety and Health considers to be potential occupational carcinogens. However, IDLH values that were originally determined in the Standards Completion Program or were recently revised are shown in brackets following the Ca designations.

[b] Not established.

IDLH, immediate danger to life and health; G, gas; MW, molecular weight.

Table 2.8 Properties of Selected Medium-Hazard Chemicals

Chemical	Structure/ Formula	MW	Odor	IDLH	Effects
Acetone cyanohydrin	$CH_3C(OH)CNCH_3$	85.1	Faint odor of bitter almond	N.D.	Eyes, skin, respiratory system, central nervous system, cardiovascular system, liver, kidneys, gastrointestinal tract
Acrolein	CH_2=CHCHO	56.1	Piercing, disagreeable odor	2	Eyes, skin, respiratory system, heart
Acrylonitrile	CH_2=CHCN	53.1	Unpleasant odor	Ca [85]	Eyes, skin, cardiovascular system, liver, kidneys, central nervous system
Allyl alcohol	CH_2=$CHCH_2OH$	58.1	Pungent, mustard-like odor	20	Eyes, skin, respiratory system
Boron tribromide	BBr_3	250.5	Sharp, irritating odor	N.D.	Eyes, skin, respiratory system
Carbon monoxide	CO	28.0	Odorless	1200	Cardiovascular system, lungs, blood, central nervous system
Ethylene dibromide	$BrCH_2CH_2Br$	187.9	Sweet odor	Ca [100]	Eyes, skin, respiratory system, liver, kidneys, reproductive system
Hydrogen selenide	H_2Se	81.0	Resembling decayed horse radish	1	Eyes, respiratory system, liver
Methyl bromide	CH_3Br	95.0	Chloroform-like odor at high concentrations	Ca [250]	Eyes, skin, respiratory system, central nervous system
Methyl hydrazine	CH_3NHNH_2	46.1	Ammonia-like odor	Ca [20]	Eyes, skin, respiratory system, central nervous system, liver, blood, cardiovascular system
Methyl isocyanate	CH_3NCO	57.1	Sharp, pungent odor	3	Eyes, skin, respiratory system

(continued)

Table 2.8 Properties of Selected Medium-Hazard Chemicals (Continued)

Chemical	Structure/ Formula	MW	Odor	IDLH	Effects
Methyl mercaptan	CH_3SH	48.1	Disagreeable odor like garlic or rotten cabbage	150	Eyes, skin, respiratory system, central nervous system, blood
Nitrogen dioxide	NO_2	46.0	Pungent, acrid odor	20	Eyes, respiratory system, cardiovascular system
Phosphine	PH_3	34	Fish- or garlic-like odor	50	Respiratory system
Phosphorus oxychloride	$POCl_3$	153.3	Pungent and musty odor	N.D.	Eyes, skin, respiratory system, central nervous system, kidneys
Selenium hexafluoride	SeF_6	193.0	Pungent	2	Respiratory system
Stibine	SbH_3	124.8	Disagreeable odor like hydrogen sulfide	5	Blood, liver, kidneys, respiratory system
Sulfuryl fluoride	SO_2F_2	102.1	Odorless	200	Eyes, skin, respiratory system, central nervous system, kidneys
Tellurium hexafluoride	TeF_6	241.6	Repulsive odor	1	Respiratory system
n-Octyl mercaptan	$CH_3(CH_2)_7SH$	146.3	Mild odor	NE[a]	Eyes, skin, respiratory system, blood, central nervous system
Other TICs in medium-hazard category	Allylamine, allyl chlorocarbonate, carbonyl sulfide, chloroacetone, chloroacetonitrile, chlorosulfonic acid, diketene, 1,2-dimethylhydrazine, methanesulfonyl chloride, methyl chloroformate, methyl chlorosilane, phosphorus pentafluoride, silicon tetrafluoride, sulfur trioxide, sulfuryl chloride, titanium tetrachloride, trichloroactyl chloride, trifluoroacetyl chloride				

[a] Not established.
IDLH, immediate danger to life and health.

Table 2.9 Low-Hazard TICs

Chemical Name	Chemical Name
Allyl isothiocyanate	Hexachlorocyclopentadiene
Arsenic trichloride	Hydrogen iodide
Bromine	Iron pentacarbonyl
Bromine chloride	Isobutyl chloroformate
Bromine pentafluoride	Isopropyl chloroformate
Bromine trifluoride	Isopropyl isocyanate
Carbonyl fluoride	n-Butyl chloroformate
Chlorine pentafluoride	n-Butyl isocyanate
Chlorine trifluoride	Nitric oxide
Chloroacetaldehyde	n-Propyl chloroformate
Chloroacetyl chloride	Parathion
Crotonaldehyde	Perchloromethyl mercaptan
Dimethyl sulfate	sec-Butyl chloroformate
Diphenylmethane-4,4'-diisocyanate	tert-Butyl isocyanate
Ethyl chloroformate	Tetraethyl lead
Ethyl chlorothioformate	Tetraethyl pyroposphate
Ethyl phosphonic dichloride	Tetramethyl lead
Ethyl phosphonothioic dichloride	Toluene 2,4-diisocyanate
Ethyleneimine	Toluene 2,6-diisocyanate

CHAPTER 3

Governmental Policies and Programs

U.S. policy is "no first use" of lethal or incapacitating chemical agents by the U.S. Armed Forces. However, the right is reserved to retaliate, using lethal or incapacitating chemical agents, against an enemy that has used them on U.S. forces. The authority to order or approve the first retaliatory usage rests with the president. Risk to civilian populations is to be avoided to the maximum extent possible.

As discussed in Chapter 2, CWAs and TICs will produce harmful physiological and/or psychological effects after exposure through inhalation, direct contact, or ingestion. Because these CWAs can cause large number of casualties, the government has developed many regulations in an effort to control, safeguard, and provide safe handling protocols. For example, chemical agent workers must meet stringent qualification criteria established in the Chemical Personnel Reliability Program to qualify for working with the agents. Pertinent regulations have been developed by the Department of Defense (DoD), Department of the Army, Army Materiel Command, and Aberdeen Proving Ground. These regulations complement each other to cover multiple aspects of safely handling CWAs. In addition to complying with diverse regulations, laboratories must develop standing operational procedures (SOPs) for each operation involving CWAs. SOPs are discussed in Chapter 4.

CWAs and TICs can be disseminated using a variety of methods and available distribution systems. Incidents in places where large numbers of people congregate or travel through, such as subway systems, airport terminals, or high-rise office buildings, could lead to hundreds or thousands of casualties and deaths (direct and indirect) and the disruption of crucial services. The U.S. government has developed numerous initiatives to deal with such incidents as part of the Domestic Preparedness Program (DPP).

The DPP includes training and equipping local response teams, such as police, fire fighters, first responders, and medical support people. In 1996, this program took the initiative to test and evaluate the detection equipment that emergency responders were using to determine if they had the CWA detection capability. The program later invited detector developers to submit products for evaluation. Evaluation results are posted on the Homeland Defense website (http://www2.sbccom.army.mil/hld/ip/reports.htm#detectors). This initiative assessed current detector capabilities as well as responders' degree of readiness. The detector evaluation program followed evaluation criteria similar to those applied in testing military CWA detection devices.

3.1 CWA DETECTION STANDARDS AND CRITERIA FOR DEPLOYMENT

Government recommendations for CWA concentration levels to be considered were prepared to serve as guidelines for the developers and researchers to meet. The recommendations are based on evaluation of previous standards to address a myriad of unique scenarios. Department of the Army pamphlet 385-61 entitled "Toxic Chemical Agent Safety Standards" details standard procedures and protective equipment for proper handling of CWAs.

3.1.1 Low-Level Exposure and Operational Risk Management

DoD policy and doctrine are based on a uniformly defined "range" of low-level exposure. This range must address all military scenarios to appropriately determine policy, doctrine, research, and technological needs. Policy and doctrine should not arbitrarily dictate either the number or percentage of casualties that a commander can or should accept in order to complete a mission. The specific accepted risk should be determined by the commander(s), based on the situation and mission requirements. Operational risk management (ORM) is a fundamental aspect of military decision making. The U.S. Army Center for Health Promotion and Preventive Medicine (USACHPPM) recommends using the ORM framework to define low-level exposures, and thus low-risk exposures.

The ORM process involves ranking the severity of a situation, and the probability that this exposure will occur. As Table 3.1 shows, low risk can be attributed to hazards ranging in severity from "negligible" to "marginal" to "critical," depending on the likelihood that the hazard will occur.

Table 3.1 Risk Assessment Matrix

	Probability				
Severity	**Frequent**	**Likely**	**Occasional**	**Seldom**	**Unlikely**
Catastrophic	E	E	H	H	M
Critical	E	H	H	M	L
Marginal	H	M	M	L	L
Negligible	M	L	L	L	L

E, extremely high risk; H, high risk; M, moderate risk; L, low risk.

Source: U.S. Department of the Army. USACHPPM Technical Guild 230, Chemical Exposure Guidelines for Deployed Military Personnel Version 1.3 May 2003; currently being reviewed by NRC.

The severity of health effects from chemical exposure is the direct result of dose, the product of chemical concentration and duration of exposure. It is impossible to predict responses and outcomes in all situations due to variation in environmental and exposure conditions. Therefore, generic scenarios are used to represent typical conditions for estimating the severity of effects.

3.1.1.1 Significant Adverse Effect

The "significance" of an effect depends on the severity of resulting health and operational impacts. While the threshold for lethal effects is defined as the high end of low-level exposures, the lower end can then be defined as the exposure dosage that would not cause significant immediate or delayed adverse health consequence.

To determine an exposure level that does not result in adverse health effects is complicated by the nature of the pathologic sequelae following exposures. If a temporary chemical exposure results in an immediate effect, the relationship between the exposure level and the physiologic impairment may be readily determined. A dose–response relationship could then be established. However, single or multiple exposures to a toxicant may cause cellular damage that is not reflected as overt pathology for days, weeks, or years following the exposure. Such pathologic sequelae are difficult to link quantitatively to a specific toxicant exposure.

Under circumstances where the physiologic damage is reversible, the resulting adverse health consequences will also reverse at some time after the exposure. This leads to observations of immediate, transient adverse health effects and/or delayed, transient adverse health effects. If sufficient transient damage accumulates to cause irreversible deterioration or if the cellular pathologic effects of exposure are irreversible, temporary or short-term exposure may result in permanent chronic illness. Therefore, adverse health effects following a chemical exposure may be a mixture of immediate, delayed, transient, or chronic symptoms. Each symptom may be characteristically associated with a different exposure level.

3.1.1.2 Duration of Exposure

The time period that personnel may be exposed to CWAs or other hazards cannot be precisely estimated. Duration periods listed in Table 3.2 describe generalized

Table 3.2 Designation of Reference Exposure Duration for Deployments

Temporary exposure duration	Brief, one-time occurrence. Such occurrence may only last minutes or up to a few hours.
Short-term exposure duration	In general, this term applies to exposures that exceed the "temporary" duration and continue daily up to 2 weeks. This includes continuous exposures and repeated, intermittent exposures.
Long-term exposure duration	Long-term exposures include continuous exposures or repeated, intermittent exposures that continue daily for more than 2 weeks.

Source: U.S. Department of the Army (see p. 34).

duration of exposure that deployed forces may encounter. This grouping provides a systematic means of focusing research. For example, the primary focus of research will be to identify low-level concentrations and associated effects for temporary exposures, as this is the most probable exposure duration anticipated. Long-term exposures to chemical agents, however, are relatively unlikely, and thus of lower priority for research. The time period of interest used by planners at the U.S. Army Center for Health Promotion and Preventive Medicine (Aberdeen Proving Ground, Maryland) is generally between 6 and 48 hr. As previously noted, duration of exposure will be variable for almost any scenario. One area of research interest is to further investigate the relationship of the concentration and time (Ct) variables to provide more accurate information to the operational and developer community.

3.1.1.3 Low-Level Exposure Concentration

A low-level exposure represents the level(s) below which there is low risk to human health. For purposes of evaluating and establishing DoD policies, doctrine, and research needs, "benchmark" concentrations are developed in accordance with the best available scientific data and risk assessment techniques. A methodology has been developed by a national advisory committee, which was established under Public Law 92-463 for interpreting toxicological data and deriving acute human exposure concentrations, referred to as acute exposure guideline levels (AEGLs). The committee is responsible for developing AEGLs for TICs. AEGLs are designed to make protective action decisions for the general public in the event of chemical releases — accidental occurrences or intentional terrorist acts. AEGLs reflect temporary exposure duration (10 min to 8 hr) for three levels of severity that are compatible with the range of military low-risk operational definitions (Table 3.3).

3.1.2 Uncertainties in Risk Assessment and Research Considerations

A key challenge for the scientific community is to develop a valid methodology and appropriate toxicological principles for predicting CWA dose–response effects over longer exposure times and at lower concentrations. A valid methodology requires developing techniques that are verifiable and defensible aimed at providing consistent and accurately measuring agents in a test chamber. To date, researchers have encountered technical challenges in generating a constant low level of chemical agents for long exposure periods and developing sampling and analysis methods to verify low-level exposure concentrations within a test chamber throughout the exposure period. Table 3.4 describes several risk assessment uncertainties that will have to be addressed when applying laboratory data to real-world scenarios.

3.1.3 Summary of Existing/Recently Proposed Air Standards

Tables 3.5 and 3.6 summarize the current status of various existing standards and provide the concentration levels and their designated application by type of population and designated time period for which the concentration applies. Very simply, the shorter the duration, the higher the concentration of a given substance

Table 3.3 Range of Low-Level Concentration and Corresponding Effects

Critical (C), AEGL-3

Above critical concentration level:
Significant fatalities (this is above low-level range).

At critical concentration:
Defines threshold for fatalities (few personnel impacted this severely, immediate effects). High end of "low-level" range.

Between critical and marginal concentration levels:
Severe immediate effects (functional decrement, may require medical attention); and/or severe transient effects, and/or probable concern for permanent/chronic health effects.

Marginal (M), AEGL-3

At marginal concentration level:
Threshold for which initial cases of severe and/or permanent health effects may occur in exposed population.

Between marginal and negligible concentration levels:
Some immediate/transient effects with mild functional decrement, limited medical attention required, and/or possible concern for permanent/chronic health effects.

Negligible (N), AEGL-1

At negligible concentration level:
Threshold for where initial signs/symptoms are initially anticipated in some members of exposed population.

Below the negligible concentration level indicates no anticipated biologically significant effects (below low-level exposure range). Biological markers for exposure are not by themselves considered significant effects.

Source: U.S. Department of the Army (see p. 34).

Table 3.4 Laboratory vs. Battlefield Exposure Conditions

Conditions	Laboratory	Battlefield
Weather (wind, humidity, temperature, air stability, precipitation)	Controlled, simple	Uncontrolled, complex
Sunlight	Artificial or none	Present (variable)
Topography	Fixed, simple	Variable, complex
Vegetation/soil	None	Varieties present
Chemical state	Vapors, aerosols, liquids (pure)	Vapors, aerosols, and liquids (likely to be contaminated)
Rate of exposure	Intermittent, continuous, sustained	Variable and intermittent
Personnel	None; animal models used	Personnel exposed and stressed (physiologically and psychologically)

Source: U.S. Department of the Army (see p. 34).

Table 3.5 Current Proposed Exposure Standards for General Population

Standard	Population	Exposure	H/HD/HT	GA	GB	GD/GF	VX	Lewisite
IDLH (immediately dangerous to life or health)	Civilian/DoD worker	One-time exposure	2	0.1	0.1	0.05	0.01	0.003
STEL (short-term exposure limit)	Civilian/DoD worker	Occasional 15-min exposure (4 times/day)	0.0004	0.0004	0.0004	0.0002	0.00004	NA
WPL (worker population limit)	Civilian/DoD worker	8 hr/day for 30 years, time-weighted average	0.003	0.0001	0.0001	0.00003	0.00001	0.003
GPL (general population limit)	Civilian population	24 hr/day, lifetime, time-weighted average	0.00002	0.000003	0.000003	0.000001	0.0000003	0.0001

NA, not available. Concentrations are listed as mg/m^3.

Source: U.S. Department of the Army (see p. 34).

Table 3.6 Proposed AEGLs: One-Time Exposures for Emergency Planning

Duration of One-time Exposure	H/HD/HT (mg/m³)	GA (mg/m³)	GB (mg/m³)	GD/GF (mg/m³)	VX (mg/m³)
AEGL, Level 1 (Nonsignificant but possibly noticeable effects)					
10 min	0.4	(0.0069)	(0.0069)	(0.0035)	(0.001)
30 min	0.13	(0.0040)	(0.0040)	(0.0020)	(0.00033)
1 hr	0.067	(0.0028)	(0.0028)	(0.0014)	(0.00017)
4 hr	0.017	(0.0014)	(0.0014)	(0.0007)	(0.000041)
8 hr	0.008	(0.0010)	(0.0010)	(0.0005)	(0.000021)
AEGL, Level 2 (Could cause casualties)					
10 min	0.6	(0.087)	(0.087)	((0.044)	(0.015)
30 min	0.2	(0.05)	(0.05)	(0.025)	(0.0050)
1 hr	0.1	(0.035)	(0.035)	(0.018)	(0.0025)
4 hr	0.025	(0.017)	(0.017)	(0.0085)	(0.00063)
8 hr	0.013	(0.013)	(0.013)	(0.00065)	(0.00031)
AEGL, Level 3 (Could cause fatalities)					
10 min	6.1	(0.76)	(0.38)	(0.38)	(0.38)
30 min	4.2	(0.38)	(0.19)	(0.19)	(0.19)
1 hr	2.1	(0.26)	(0.13)	(0.13)	(0.13)
4 hr	0.53	(0.14)	(0.070)	(0.070)	(0.070)
8 hr	0.27	(0.010)	(0.051)	(0.051)	(0.051)

Note: Values in parentheses are taken from draft documents not yet published.
AEGL, acute emergency guideline level.
Source: U.S. Department of the Army (see p. 34).

that a person can tolerate. If repeated exposures over a long period of time are anticipated, the resulting daily average dosage will be lower than the stated average over a short period of time. AEGL levels are most pertinent to military applications.

3.1.4 Recommended Chemical Agent Concentration Criteria for Detectors

There are both scientific and political reasons that justify reassessment of current criteria. The USACHPPM has been careful to focus on realistic operational scenarios in this assessment. Current military detector capabilities may not immediately be able to meet all levels of the expanding range of what may constitute a "low-level" risk during military operations. Current equipment capabilities are generally adequate to identify and thus allow decision making regarding chemical agent exposures at the high end of the low-risk range. The lower end of that range, which would include conservative estimates of the threshold for significant biological effects in the general population, is not yet achievable for all agents.

Therefore, the 10-min AEGL Level-1 concentration values for each agent should be established as alarm trigger points in developing new detectors. These concentrations should be detected in 10 min or less. The proposed capabilities of equipment currently in development (e.g., the U.S. military's Joint Chemical Agent Detector, a multimission chemical agent point detection system [JCAD]) aim to be near the low extreme of the low-level concentration range.

For equipment that is designated specifically for wartime use, setting alarms at higher thresholds may be warranted. The use of AEGL-2 or AEGL-3 values should be considered as thresholds and objectives. Future detectors should be able to continuously monitor airborne concentrations down to the designated 8-hr AEGL concentration. Table 3.7 lists capabilities and objectives of selected existing and development-stage detectors. The rationales for these recommended criteria include the following:

1. The values are scientifically defensible. These AEGLs are based on a scientifically acceptable method currently used for assessing other TICs.
2. The AEGLs are derived for scenarios that parallel military deployment scenarios. Exposure durations address the highest-probability exposure scenarios of military interest. Anticipated scenarios where military troops may be exposed to CWAs include a rare, if not one-time brief event, where exposure would last minutes to a maximum of hours. The three AEGL levels provide the necessary flexibility for military ORM decision making. Use of AEGLs prevents establishing a single, conservative "acceptable level" for each chemical, such as those designed for chronic lifetime exposure periods that are unlikely in military applications.
3. These values are protective of the general population. The AEGLs are designed for the general public, and thus are an appropriate way to ensure that more sensitive members of the deployed military population are being protected.

3.2 JOINT SERVICES OPERATIONAL REQUIREMENTS FOR CHEMICAL AGENT DETECTORS

The purpose of the Joint Services Operational Requirements (JSOR) (Table 3.8) for CWA detectors is to permit timely warning for people to don protective gear. Criteria for the Automated Chemical Agent Detection Alarm (ACADA), in use by the U.S. military, are based on the JSORs, which were established without considering airborne exposure limits (AELs). The JSOR values consider adequate advance warning time to escape the area without significant harm when the air contains a particular concentration of a target chemical. Detection technologies have not yet been developed to permit detection of lowest-level concentrations under the AEL criteria.

Another set of criteria for the Joint Chemical Agent Detector (JCAD) has been proposed. The JCAD includes the ACADA JSOR requirements. Detector sensitivity down to the AEL is being considered as technology progresses.

Table 3.7 Existing Detector Capabilities and Proposed Thresholds and Objectives Compared to AEGLs

Detector	Agent						
	GA	GB	GD	GF	VX	HD	Lewisite
Minicams/RTAP	—	0.0001	0.00003	0.00003	0.00001	0.00003	0.0006
ACAMs	—	0.0001	—	—	0.00001	0.001	—
DAAMs	—	0.0001	—	—	0.00001	0.001	—
Bubbler	—	0.0001	—	—	0.000001	0.003	0.003
CAM	0.03	0.03	0.03	0.03	0.01	0.1	2.0
Draeger	—	—	—	—	3–2.5	—	—
ICAD	0.5	0.5	0.5	—	—	10.0	10.0
M18A2	—	0.02	—	—	0.1	0.5	10.0
M21	3.0	3.0	3.0	—	—	150	150
M90	0.04	0.04	0.04	0.04	0.04	0.2	0.2
M256	—	0.05 mg/m³	—	—	0.1	1–3	14
M256A1	—	0.005 mg/m³	—	—	0.02	3.0	14
JCAD (Proposed)	0.001; <30 min (T); <15 min (O)	0.001; <30 min (T); <15 min (O); 0.001 mg/m³ in <30 min (T); <15 min (O)	0.001; <30 min (T); <15 min (O)	0.001; <30 min (T); <15 min (O)		0.02; <30 min (T); 0.003 mg/m³ in 15 min (O)	0.02; <30 min (T); 0.003 mg/m³ in 15 min (O)

Note: Concentration values are listed as milligrams per cubic meter.

AEGL, acute emergency guideline level; O, objective; T, threshold.

Source: U.S. Department of the Army. Army Regulation AR-385-61: The Army Chemical Agent Safety Program, Safety, 28 February 1997.

Table 3.8 Joint Services Operational Requirements for Point Sampling Detectors

Agent	Threshold Exposure Concentration (mg/m³)	Threshold Exposure Response Time Maximum(s)	Relative Humidity (% RH) Range	Temperature Range (°C)
VX	1	10		
	0.04	90	5 to 100	−10 to +49
	0.1	30		
GA, GB, GD,	1	10	5 to 100	−30 to +49
GF	0.1	30		
HD, L, HN3	50	10	5 to 100	−18 to +49 (for HD)
	2	120		−18 to +49 (for L)
				+15 to +49 (for HN3)
AC	2500	10	5 to 100	−32 to +49
	22	60		
CK	20	60	5 to 100	−32 to +49

Source: U.S. Department of the Army. Proposed JCAD operational requirements Appendix A of JCAD Study Plan (Draft) 6/11/1998 modified for contractor distribution.

The JSORs are frequently mentioned in this chapter. These figures are used as guidelines for developers of modern detection instruments, and also serve as evaluation guidelines under most test programs.

3.2.1 JCAD Requirements and Rationales

The JCAD is expected to replace ACADA detectors used in the field. In order to develop a methodology for testing the JCAD, it was necessary to establish specific test temperatures and associated relative humidity levels. JCAD performance specifications are based on the aforementioned recommended criteria. The operational temperature range is from −32°C to 49°C together with relative humidity ranging from 5% to 100%.

The JCAD also has performance requirements to simultaneously detect, identify, and quantify CWA vapors at various concentration levels while in the presence of chemical interferents. This detector's performance requirements integrate dynamic agent concentration profiles and provide dosage measurements. As of this writing, a completely verified test system for evaluating JCAD performance does not exist.

To meet that goal, the government formed the Test Implementation Working Group that in turn established the Test Methodology Working Group (TMWG) to determine the methodology to be developed, validated, and verified in JCAD testing. Test methodology development (TMD) addresses chemical agent identification and detection requirements under JCAD performance specifications.

After reviewing existing and required test facility capabilities, the TMWG identified TMD in the following areas:

- Simultaneous generation of two agent vapors at two or more constant concentration levels with one or more chemical interferents at two or more concentration levels
- Simultaneous generation of two agent vapors at two or more dynamic concentration profiles with one or more chemical interferents at two or more levels
- Characterization of chemical interferents at two or more concentration levels
- Quantification of agent concentrations and dosage (integration of concentration over time) to the best available accuracy at the time of measurement with chemical interferents present
- Quantification of moving integration of agent concentrations and monitoring of hazard levels to the best available accuracy at the time of measurement with chemical interferents present
- Monitoring and recording of unit-under-test performance attribute data
- Quantification of TICs

In addition, the test system shall collect and process test data to calculate estimates of the following:

- Accuracy of measured agent concentrations and dosages
- Mean and standard deviation of exposure response times for controlled exposure agent concentration levels
- Magnitudes of any interactions and effects of controlled factors (humidity, temperature, agent type, agent concentration, dynamic profile, chemical interferent type, chemical interferent concentration) on observed performance (measurement of agent concentrations, dosage, hazard level, detection and identification response times, and time needed for reset after alerts) of the unit under test
- Probability of detection and identification of each agent or simulant as a function of agent or simulant concentration
- Probability of a false alarm and estimate of the mean time between false alarms (MTBFAs)
- Characteristics of conditions related to false alert events

3.3 OBJECTIVE AND APPROACHES AT DUGWAY PROVING GROUND

Approaches to address the above needs are included here as examples to illustrate critical requirements for proper detector evaluation. The basic technical approach to vapor generation and monitoring at Dugway Proving Ground (DPG; under the U.S. Army Test and Evaluation Command, located in Utah) is to use an evaporative disseminator for vapor generation with MINICAMS® and/or MIRAN® for "real-time" monitoring. Sorbent tubes and gas chromatography (GC) analysis are used for obtaining integrated vapor concentration measurements.

A limited amount of engineering takes place prior to each validation and verification phase. This involves determining agent–air mixtures, control settings, and instrument configurations required to achieve desired vapor concentrations. Modifications to the detector testing system (DTS) could include reconfiguring instrumentation and controls, changing dilution airflow rates, or modifying the vapor dissemination device currently in use.

JCAD performance specifications cover a wide range of agents, concentrations, temperatures, humidity, and chemical interferents. A complete matrix to test the full envelope of performance would be impractical and cost prohibitive. Therefore, the statistical experimental design is crucial for reducing the matrix without sacrificing statistical validity, while also ensuring validity of the developed test methodology. The seven TMD areas stated earlier are discussed below.

3.3.1 Simultaneous Constant Agent Vapor Concentration Generation

Objective: To generate and maintain constant vapor concentrations for two agents simultaneously. Concentrations will be at the key performance and threshold detection levels for testing. Generation of objective levels will be evaluated in follow-up testing.

Approach: The DTS is currently configured to generate key performance level concentrations, which were demonstrated individually during ACADA testing. Generating the threshold and objective concentration levels will be achieved by increasing the quantity of dilution air, reducing the surface area of evaporating agent, and/or using a diffusion-tube disseminator. Simultaneous generation of two agent vapors at key performance concentration levels is of primary concern; simultaneous generation of two agent vapors at threshold concentration levels is of secondary concern; simultaneous generation of two agent vapors at objective concentration levels is of tertiary concern. Simultaneous generation of two agent vapors will be accomplished using evaporative vapor generators or diffusion tube generators in parallel.

3.3.2 Simultaneous Dynamic Agent Vapor Concentration Generation

Objective: Simultaneously generate two agent vapors with independently varying concentration profiles. Concentration profiles will include key performance and threshold detection levels for Phase I testing; generation of objective levels will be evaluated for Phase II testing.

Approach: Dugway Proving Ground proposes to use two evaporative vapor generators operated in parallel to generate agent vapors. Agent vapor concentrations will be controlled by adjustments of dilution air feed rates.

3.3.3 Characterization of Chemical Interferents

Objective: To generate and adequately control concentration levels of chemical interferent vapors.

Approach: DPG proposes the use of an evaporative disseminator to generate chemical interference vapors. DPG proposes to use either MIRAN or other organic vapor analyzer to monitor the chemical interferent, which will enable DPG to generate chemical interferent vapors reproducibly.

3.3.4 Quantification of Dosages

Objective: The test system will quantify agent dosages independently of the item under test.

Approach: DPG will use the data from near-real-time agent monitors and supplemental sorbent tube and/or bubbler samples to quantify the dosages presented to the units under test. This data will be collected, stored in the test database, and compared to the values reported by the units under test.

3.3.5 Quantification of Hazard Levels

Objective: The test system will quantify agent hazard levels independently of the unit under test.

Approach: DPG will use data from the near-real-time agent monitors along with the algorithm for calculating hazard levels to quantify the hazard level for the agent presented to the units under test.

3.3.6 Data Monitoring and Recording

Objective: The test system will monitor and record appropriate test data and data from the units under test, and then store the data in a database. The data will be organized and reported in a fashion that will support and simplify assessment of the performance of the unit under test.

Approach: DPG will use the existing DTS data collection system to collect all test data and item-under-test data. The system will be customized for the JCAD test. The data will be stored in the existing database system. Data may also be stored in a relational database system to aid in data analysis. Data reporting formats will be developed specifically for the JCAD test project.

3.3.7 Quantification of TICs

Objective: The test fixture will be capable of generating and quantifying vapors of TICs to determine the ability of the unit under test to detect and quantify the vapor.

Approach: DPG proposes to treat the requirement for TICs as if they were additional CWAs. The test system will have the same capabilities for the TICs as for the CWAs.

3.4 JCAD REQUIREMENTS FOR DETECTION AND IDENTIFICATION FUNCTIONS

Appearing below are selected requirements for the development of future detection devices such as the JCAD. As mentioned previously, the JCAD is intended to be widely deployable to replace or complement the ACADA. The JCAD must be able to detect and identify numerous CWAs and selected TICs.

3.4.1 Detection and Identification

The JCAD shall automatically and simultaneously detect, identify, and quantify chemical vapor by class (nerve, blister, and blood, and TIC) and specific agent (GA, GB, GD, GF, VX, HD, L, HN3, AC, CK, and possibly some TICs). It shall detect and provide an alert for nerve agent vapors, blister agent vapors, and blood agent vapors within the response times at constant concentration exposures listed in Table 3.9. These requirements are to be incorporated in requirements for future detector certification. TICs' alerts are not a requirement at present. Ability to detect TICs may be desirable however.

3.4.2 Sampling Requirements and Additional Challenges

Sampling, although seemingly simple, is perhaps one of the most difficult steps in chemical and biological agent detection. Problems arise in ensuring that the sample taken represents the media sampled, which in turn depends on proper mixing. Concentrators must be developed that can circulate great quantities of air without clogging or losing effectiveness, and that can concentrate the desired molecules. Recognizing and removing environmental interference and background effects, such as those due to geographical and temperature variability, are also necessary.

Sensors and detectors require different technology according to usage. Even the smallest tabletop mass spectrometer in a standard research laboratory cannot be converted into a rugged, lightweight model that a soldier, HAZMAT team member, or first responder could carry. Room-temperature detectors; label-free detection (e.g., without fluorescent beads); and low-cost, real-time analysis of airborne particles are barriers to widespread monitoring of chemical and biological weapons in public areas. Other obstacles that need to be surmounted by chemists and chemical engineers include improved nanoscale fabrication methods, better microfluidics and macro–micro scale interfaces, better high-throughput screening, more efficient heat exchangers, and lighter batteries. Engineering challenges include integrating new technological components into a useful finished product.

To develop robust sensors, a multidisciplinary systems approach must be taken. Experimentalists, statisticians, engineers, and data analysts must be included from the beginning (conceptual stage) to the end (using the sensor in the field). Well-designed experiments are crucial in developing sensor field worthiness, with statisticians as close collaborators of chemical and biological scientists. Actual and potential interference must be identified and dealt with through hardware design, multiple sensor types, or multivariate techniques, or through software development such as statistical analysis. Sensor calibration and drift caused by environment conditions must be addressed and corrected either mathematically or through hardware design.

Once developed, detectors and sensors must be tested according to strict criteria. A well-defined and very demanding set of test standards initiated by the DoD, including limits of detection, has been in place for the last 15 years. The U.S. Army periodically offers the opportunity for realistic field testing, and wind tunnel testing is available at several sites. It is imperative that scientists take advantage of existing resources for technology validation. It would be useful to develop new, integrated

Table 3.9 Single-Dose Hazard Levels

Hazard Levels	GA	GB	GD,GF	VX	HD
None: 2–10-min exposure in mg-min/m^3	≤0.05[a]: concentration is ≤0.005 mg/m^3 for 10-min exposure; ≤0.025 mg/m^3 for 2-min exposure	≤0.05[a]: concentration is ≤0.005 mg/m^3 for 10-min exposure; ≤0.025 mg/m^3 for 2-min exposure	≤0.02[a]: concentration is ≤0.002 mg/m^3 for 10-min exposure; ≤0.01 mg/m^3 for 2-min exposure	≤0.009[a]: concentration is ≤0.0009 mg/m^3 for 10-min exposure; ≤0.0045 mg/m^3 for 2-min exposure	≤2.5[a]: concentration is ≤0.25 mg/m^3 for 10-min exposure; ≤1.25 mg/m^3 for 2-min exposure
Low: Protection from symptoms, including miosis, conjunctivitis, rhinorrhea, and tightness in chest; 2–10-min exposure in mg-min/m^3	>0.05 and <0.5: For alert — minimum concentration to detect is 0.005 mg/m^3 for 10-min exposure; maximum concentration is 0.25 mg/m^3 for 2-min exposure	>0.05 and <0.5: For alert — minimum concentration to detect is 0.005 mg/m^3 for 10-min exposure; maximum concentration is 0.25 mg/m^3 for 2-min exposure	>0.02 and <0.2: For alert — minimum concentration to detect is 0.002 mg/m^3 for 10-min exposure; maximum concentration is 0.1 mg/m^3 for 2-min exposure	>0.009 and <0.09: For alert — minimum concentration to detect is 0.0009 mg/m^3 for 10-min exposure; maximum concentration is 0.045 mg/m^3 for 2-min exposure	>2.5 and <25: For alert — minimum concentration to detect is 0.25 mg/m^3 for 10-min exposure; maximum concentration is 12.5 mg/m^3 for 2-min exposure
Medium: Mask to protecting lungs; 2–10-min exposure in mg-min/m^3	≥0.5: For alert — concentration is >0.05 mg/m^3 for 10-min exposure; >0.25 mg/m^3 for 2-min exposure	≥0.5: For alert — concentration is >0.05 mg/m^3 for 10-min exposure; >0.25 mg/m^3 for 2-min exposure	≥0.2: For alert — concentration is >0.02 mg/m^3 for 10-min exposure; >0.1 mg/m^3 for 2-min exposure	≥0.09: For alert — concentration is >0.009 mg/m^3 for 10-min exposure; >0.045 mg/m^3 for 2-min exposure	≥25: For alert — concentration is >2.5 mg/m^3 for 10-min exposure; >12.5 mg/m^3 for 2-min exposure
High: Suit protecting skin; to 30–50-min exposure in mg-min/m^3	≥1000: For alert — concentration is ≥20 mg/m^3 for 50-min exposure; ≥33 mg/m^3 for 30-min exposure	≥600: For alert — concentration is ≥12 mg/m^3 for 50-min exposure; ≥20 mg/m^3 for 30-min exposure	≥150: For alert — concentration is ≥3 mg/m^3 for 50-min exposure; ≥5 mg/m^3 for 30-min exposure	≥5: For alert — concentration is ≥0.1 mg/m^3 for 50-min exposure; ≥0.17 mg/m^3 for 30-min exposure	≥25: For alert — concentration is ≥0.5 mg/m^3 for 50-min exposure; ≥0.83 mg/m^3 for 30-min exposure

[a] No observable adverse effect level.

Source: U.S. Department of the Army. JCAD Study Plan for Test Methodology Development Modified for Contractor Distribution, Final Draft 6/11/1998.

multiple-source databases to create chemical and biological agent libraries for quick agent identification and access to neutralization methods. Libraries already exist in many individual agencies, and there is a need for consolidation. Sharing existed data among various agencies would prevent repetitive expenditures to seek the same information. Consolidation of data sharing is most cost effective to free funds for additional researches.

3.5 GENERAL CAPABILITIES NECESSARY TO MITIGATE VULNERABILITY

Capabilities of existing commercial sensors are limited. Most manufactured sensors are designed for use in specific environments to detect compounds of their interest. Sensor systems that can detect a large number of chemicals are needed at present. Given the high toxicity of the CWAs and TICs, detection sensitivity of many existing sensors must be improved significantly.

Sensors to assist authorities in determining when a site is safe for normal functions to resume are required for postincident management. For that to happen, detection sensors must have significantly increased specificity and sensitivity beyond any currently available systems.

Consequently, various subsystems will also be needed to support the development of these more sensitive instruments. Systems for reliable sample collection, sample processing, and presentation of chemicals to sensors are essential. Standardized methodologies are needed. Systematic quality assurance of sensor evaluations can only be achieved through standard methodologies that have proved successful. Toward this end, the National Institute of Standard Technologies (NIST) is in the process of generating the standard testing protocol under which all future detection devices shall be tested for certification.

The U.S. government supports research of these sensors mainly through the DoD, National Science Foundation, and Department of Energy. Sensor development is also heavily supported by private industry. New sensors or technology improvements are hitting the market fairly rapidly through these research efforts. However, none of the current technologies has had any real impact on emergency preparedness. For sensors to be effectively implemented, they will have to be fairly inexpensive, widely deployable, and networkable.

Development of sensors that can detect and identify the release of toxic materials must continue. Effective responses to the specific agent involved in a chemical attack can only be achieved through the correct choice of sensors for the job, which in turn is crucial for effective consequential management including orderly evacuation to minimize casualties. Therefore, a program with sustained funding to focus and coordinate research and development on sensors and sensor networks, with an emphasis on fielding systems, is needed.

The following sections provide an example of how a detection device is tested in general to qualify as a viable device. The protocol was developed specifically for the DPP for detector evaluation. Although the protocol is limited in scope, it has demonstrated effectiveness in general characterizations of devices tested thus far. It has been well accepted, and therefore has been proposed to the NIST to serve as a platform for its detector testing protocol.

3.6 EVALUATION OF COMMERCIALLY AVAILABLE DETECTION DEVICES FOR CERTIFICATION AS CWA DETECTORS

3.6.1 Background

In 1996, the DoD established the DPP. One of its objectives is to enhance federal, state, and local emergency responders' capabilities to respond to nuclear, biological and chemical terrorism incidents. Emergency responders who encounter a potentially contaminated area must survey the area for the presence of toxic compounds, including CWAs and explosives. Vapor content detectors commonly used or commercially available must be evaluated for their ability to effectively detect and identify CWAs.

Under the DPP Expert Assistance (Test Equipment) Program, the SBCCOM established a program to address this need. The Applied Chemistry Team, formerly known as the Design Evaluation Laboratory at Aberdeen Proving Ground in Edgewood, Maryland, has been conducting detector testing for the DPP for several years. The Applied Chemistry Team is tasked with providing the necessary information to aid authorities in selecting detection equipment applicable to specific needs.

The Department of Commerce, NIST's Office of Law Enforcement Standards (OLES), and SBCCOM amended Interagency Agreement MO2398 to develop standards for testing chemical and biological detection and protective equipment for first responders. The task includes validating test methods through testing representative items, documentation of test procedures, and development of a laboratory consortium with the OLES/NIST for chemical and biological detector certification testing.

3.6.2 Proposal

A proposed protocol submitted to the NIST committee for consideration is described below. Most, if not all, elements of the protocol are expected to become requirements for future certification as a viable CWA detection device. Techniques described in this protocol serve as approaches that laboratories may follow. While suggested methodologies may not be the only or the best that are available, they have been thoroughly employed and have demonstrated highly satisfactory results for many years. Any deviation from these procedures must be thoroughly proven valid before use. That is the procedure must not cause artifacts. The ultimate objective is to evaluate the test item with the following questions in mind:

- Can the detector establish how clean the environment is? Can the measurements provide both concentration and dosage to verify concentration and degree of hazard?
- For missions of various durations, what level of decontamination is desired and does the detector provide capability to these levels?

The assessment includes identification of specifications and requirements that could improve the description of necessary detection capabilities.

3.6.3 Purpose

Techniques in CWA vapor generation, sample analytical procedures, testing procedures, and documented target detection and identification sensitivities are to serve as minimum guidelines for laboratories that conduct certification testing of detection devices. The DPP evaluation protocol is designed to characterize the CWA vapor detection capability of the candidate detector. Because covering all CWAs is impractical, agents used in testing include tabun (GA), sarin (GB), and mustard (HD) because they are believed to be the most likely threats. The protocol is limited to permit screening of devices within reasonable time and budget constraints.

Although the above three agents are to be tested only during initial screening, the candidate device must be fully programmed to include detection algorithms for all other agents that it is intended to detect. Testing for other agents will be added depending on performance of the device in the initial screening.

Test procedures and concepts used in the DPP program follow:

1. For each selected CWA, determine the minimum concentration levels (minimum detectable levels, MDLs) where repeatable readings are achieved using JSORs for point sampling devices as a guide for detection sensitivity objectives. The MDL is determined at ambient temperature and 50% relative humidity.
2. Determine the detector response for each CWA at ambient temperature at various RH levels and temperature extremes to investigate the effects of humidity and temperature on detector behaviors.
3. Observe the effects of potential interference with commonly found substance vapors to assess the ability of the detector to perform both in the laboratory and in the field.

3.6.4 General Test Protocol

A candidate device is evaluated according to the following minimum criteria.

3.6.4.1 Operating Characteristics

Startup characteristics and time required for the detection device to achieving readiness for reliable operation in various environments are observed. Total time to readiness (reliable performance) must be reasonably short. Detectors must be relatively easy to operate.

3.6.4.2 Device Sensitivity

Each device is challenged with concentrations of the CWA/compound vapor to determine its threshold sensitivity using a vapor generated under controlled atmospheric conditions. The threshold concentration is defined as the concentration of the substance at which an alarm within 2 min of exposure time or a definitive detection is observed. CWA vapor concentrations are determined by appropriate analytical methods.

The determined threshold concentration level or minimum detectable level (i.e., MDL) is used in subsequent tests at various humidity and temperature conditions. MDL is determined based on detection occurrence with a minimal target substance dosage. For example, a longer exposure time is permissible if the device can demonstrate capacity to detect the target substance at a lower concentration to produce the lowest concentration/time relationship (Ct).

Testing the device with higher vapor concentration levels produces information on its ability to quickly provide the necessary warning to minimize exposure dosage and to clear out vapor residue after contamination. Detectors are tested at low, medium, and high concentration levels to generate response curves.

Current JSOR point-detection target sensitivities for the JCAD, as shown in Tables 3.9 through 3.13, serve as guides for candidate detector tests to follow. These targeted concentrations were developed according to the relative toxicity of each CWA and the current state-of-the-art in detectors. JCAD requirements are extensions of the JSOR requirements for the ACADA; JCAD includes even lower concentration detection, with AEL detection as an ultimate goal.

Tables 3.9 through 3.13 provide the required detection performance of devices designed for a specific sensitivity range. Table 3.9 shows the exposure dosage that will cause physiological effects. IDLH concentrations and AELs serve as guides for appropriate protective equipment requirements. AEL values are currently considered permissible exposure levels for up to 8 hr without special protection. IDLH levels for GA and GB are 0.2 mg/m^3 (0.030 ppm) and 0.2 mg/m^3 (0.035 ppm), respectively. An IDLH for HD has not been established due to concerns about carcinogenicity. AELs for GA, GB, and HD for persons without masks are 0.0001 mg/m^3 (0.00002 ppm), 0.0001 mg/m^3 (0.00002 ppm), and HD (0.003 mg/m^3 or 0.0005 ppm), respectively.

Table 3.10 Maximum Response Time at IDLH and Higher Concentrations

Agent	Threshold Exposure Concentration (mg/m^3)	Threshold Exposure Response Time Maximum(s)	Relative Humidity (% RH) Range	Temperature Range (°C)
VX	1	10		
	0.04	90	5 to 100	−10 to +49
	0.1	30		
GA, GB, GD,	1	10	5 to 100	−30 to +49
GF	0.1	30		
HD, L, HN3	50	10	5 to 100	−18 to +49 (for HD)
	2	120		−18 to +49 (for L)
				+15 to +49 (for HN3)
AC	2500	10	5 to 100	−32 to +49
	22	60		
CK	20	60	5 to 100	−32 to +49

Source: U.S. Department of the Army (see p. 42). JCAD Study Plan (draft 5/11/94).

Table 3.11 Maximum Response Time at Low Concentrations

Agent	Threshold Exposure Concentration (mg/m^3)	Threshold Exposure Response Time Maximum(s)	Relative Humidity (% RH) Range	Temperature Range (°C)
VX	0.001	1800	5 to 100	−10 to +49
GA, GB, GD, GF	0.001	1800	5 to 100	−30 to +49
HD, L, HN3	0.02	1800	5 to 100	+15 to +49
AC	NA	NA	5 to 100	−32 to +49
CK	NA	NA	5 to 100	−32 to +49

NA, not available. To be determined.

Source: U.S. Department of the Army. Appendix A of JCAD Requirements Study Plan modified for contractor distribution (draft 6/11/98).

Table 3.12 Objective Maximum Response Time at AEL Concentrations of Selected Agents

Agent	Exposure Concentration (mg/m^3)	Exposure Response Time Maximum(s)	Relative Humidity (% RH) Range	Temperature Range (°C)
VX	0.00001	900	5 to 100	−10 to +49
GA, GB, GD, GF	0.0001	900	5 to 100	−30 to +49
HD, L, HN3	0.003	900	5 to 100	+15 to +49
AC	NA	NA	5 to 100	−32 to +49
CK	NA	NA	5 to 100	−32 to +49

AEL, airborne exposure limits; NA, not available. To be determined.

Source:. U.S. Department of the Army. Appendix A of JCAD Requirement Study Plan modified for contractor distribution (draft 6/11/98).

Tables 3.10 through 3.12 show desired maximum detection time versus concentration levels ranging from the IDLH to targeted detection capability at AEL concentrations for respective CWAs. Table 3.13 lists the desired minimum concentration level applicable to the general public. These numbers are considered wishful thinking at this time. Efforts should be focused on the more realistic goal of reaching AEL detection and verification with reliability.

Results from certification testing will be compared to the above values to assess the relative usefulness of the detection device in general. The MDL determined places the device in its proper category.

To compare detector sensitivities, respective LDL concentrations and detection times are plotted along the vertical and horizontal axes, respectively. Constant dosage measurements are plotted along the lines determined by the following equation:

$$K_d = \text{Dosage} = \text{Concentration (C)} \times \text{Time} = \text{Ct (mg-min/m}^3\text{)} \qquad (3.1)$$

The following single-dose hazard levels are related to various observed effects on humans:

1. The concentration level that has no observed effect over a 1- to 14-day period. This is the short-term level for ocular or nasal vapor prescribed in military air

Table 3.13 Minimum Effect Concentrations of Selected Agents

Concentration (mg/m³)	GA	GB	GD	VX	HD	L
Military air guideline — short-term, ocular or nasal vapor 1- to 14-day exposure	0.00001	0.00001	0.000003	0.000003	0.003	NA
Unmasked agent worker — 8-hr time-weighted average in any work shift (demonstration standard)	0.0001	0.0001	0.00003	0.00001	0.003	0.003
No effects concentrations in any environment	0.000003	0.000003	0.000003	0.000003	NA	NA
General population, 3-day time-weighted average	0.000003	0.000003	0.000003	0.000003	0.0001	0.003

NA, not available.

Source: U.S. Department of the Army. AR 385-61 The Army Chemical Agent Safety Program, Safety, 28 February 1997.

contamination guidelines. This level is plotted as the cut-off value for longer exposure time of hazard dosage level "none." It is this hazard dosage value that detectors should measure in a given environment.

2. The concentration level that has no observed effect over an 8-hr shift in an airborne agent work environment. This level is plotted as the 8-hr time weighted average (TWA) value that the detector should measure in work environments.
3. The single-dose hazard level after a 2- to 10-min exposure in which a soldier's fighting capability is not measurably affected. This level is the upper boundary of the "none" range.
4. The single-dose hazard level after a 2- to 10-min exposure below which miosis, conjunctivitis, rhinorrhea, and tightness in chest occur if the general population is exposed to the agent. This level is the upper boundary of the "low" range.
5. The single-dose hazard level in a 2- to 10-min exposure below which 50% of exposed soldiers show effects (ECt_{50}) if eyes and lungs are not protected. This level is the upper boundary of the "medium" range. This same dosage is the lower boundary of the "high" range in which dosages during a 30- to 50-min exposure will affect 50% of the population if eyes, lungs, and skin are not protected.

To our knowledge, there is no detector at present that has the capability of measuring guideline concentrations over a 1- to 14-day period or the safe exposure level developed for the general public. No instrument is specified at this level of detection. The 1- to 14-day exposures appear to be extremely low considering that the 8-hr TWA for workers in an environment where toxic agent(s) are used is 10 to 33 times higher.

3.6.4.3 Relative Humidity and Temperature Effects

Testing of detection capability at RH levels of less than 10%, 50%, and more than 90% and at temperature extremes (determined by the candidate's operable range) using the determined threshold concentrations are done to gain insights on the device's ability to perform in various environmental conditions. Desired performance targets are also listed in Tables 3.9–3.13. Targets are current JSORs in developing the JCAD. Commercial detectors may not be able to meet military requirements. It is important to avoid exceeding the operable temperature and humidity range that the device was designed for during any evaluation.

3.6.4.4 Field Interference Test

The ability to detect a substance in the laboratory under clean, controlled conditions is insufficient to determine whether a detection device is useful. The units need to be tested in the field with common potential interference substances such as engine exhaust, burning fuels, and other burning materials such as common clothing and building materials. Such tests reveal potential problems under real-world operational conditions. Table 3.14 lists potential field interference substances tested.

Because testing CWAs in the open air is not possible, potential interfering substances are tested in the laboratory at controlled exposure levels. The detector's ability to detect the CWA vapor is tested in combination with potential interfering vapors at the 0.1% and 1% headspace saturation concentration level of interferent vapor, providing that the interferent does not cause the detector to alarm. Such testing reveals whether the detector issues false positive or false negative results.

Substance vapors that are considered potential interferents should be screened using a controlled generation technique. The headspace vapor of the substance is blended with an airstream to produce approximately 1% concentration of the interferents. If a false alarm occurs, the concentration is lowered to approximately 0.1% and retested. If a detector does not respond to the interference vapor, then the airstream is replaced with a stream of similar air that contains the target CWA in order to assess detection of the target under the influence of the interference vapor.

Even if a particular detector has CWA detection and identification ability, its usefulness is severely diminished if it shows a similar response to too many other substances. Table 3.15 lists common potential interference substances to be tested. Others can be added as necessary.

Table 3.14 List of Potential Field Interferents to be Tested Outdoors

Gasoline exhaust, idle	Insect repellent
Gasoline Exhaust, revved	Diesel vapor
Diesel exhaust, idle	Gasoline vapor
Diesel exhaust, revved	HTH (supertropical bleach, 107, calcium hypochloride) vapor
Kerosene vapor	Bleach vapor
Kerosene on fire	Burning cardboard
JP8 (jet fuel) vapor	Burning cotton
Burning JP8 smoke	Burning woodfire smoke
Burning gasoline smoke	Doused woodfire smoke
Burning diesel smoke	Burning rubber
AFFF (Aqueous Film Forming Foam) vapor	

Source: U.S. Department of the Army. Domestic Preparedness Program. Protocol for Detector Testing and Evaluation.

Table 3.15 Typical Substances Tested in Laboratory as Interferents

AFFF (Aqueous Film Forming Foam)	
Antifreeze	Vinegar
Household bleach	Windex™
Diesel fuel	Floor wax
Gasoline	Spray 9 cleaner™
Jet fuel-JP8	25-ppm ammonia
Toluene	

Source: U.S. Department of the Army. Longworth et al. Domestic Preparedness Program: Report on Testing of Commercially Available Detectors against Chemical Warfare Agents. (http://hldisbccom.army.mil).

3.6.5 Stability and Reliability

General observations made during the test program are sufficient to determine the relative stability and reliability of detector devices. Abnormalities and problems are recorded. No special stability or reliability runs are necessary during this preliminary testing. The stability or reliability of the instrument will be determined only after the instrument has demonstrated potential as a CWA detection device.

3.6.6 Remarks

The extent of this type of testing is limited to the abovementioned characterization criteria. Test conditions should not exceed a device's potential as provided by the manufacturer. The evaluation protocol is intended to provide an abbreviated and yet sufficient characterization of candidate devices to assess CWA detection capabilities to aid authorities in selecting equipment for CWA detection.

The scope of this type of test program suggests that statistical analysis is not necessary because each test point will be conducted with repetitions using multiple units. The comparative data collected will provide insight on device performance.

Of primary importance is the ability to detect the target CWA within acceptable time and concentration limits to protect the users such as first responders. The alarm threshold must be able to provide sufficient time for responders to don protective gear before they become casualties. Failure to meet this criterion is grounds for termination of further testing.

Also important is the ability to resist false alarms. False alarms could be expected in high-sensitivity detectors no matter what technologies are used. The key is to assess false alarm rates and the materials that caused the false alarms. If the device responds with false alarms too frequently when exposed to common substances, its usefulness is greatly diminished. Equally if not more important is the ability of a device to function properly while under the influence of interferent vapors. If a relatively large number of interferent vapors affect a device's ability to detect the target chemicals, its usefulness is also limited.

The tested device must be easy to operate and require minimal recalibration to produce reliable results after a period of storage. The ability to achieve operational stability within a short time after setup is an important criterion for usefulness to first responders in emergency situations.

Testing should be divided into phases. The first phase screens the device for CWA sensitivity at ambient temperature and humidity conditions after a successful startup procedure. The dosage derived from the product of substance concentration and alarm time determines whether the testing should continue to the next phase.

The second phase includes testing the device in various temperature and humidity conditions. Successful completion would then lead to additional testing with potential interference substances both outdoors in the field and in the controlled laboratory environment. Given successful results in the second phase, the testing should then be expanded to cover other agents and other features of the device.

3.6.7 CWA Sensitivity Testing

Agent vapor generation methodologies are described in "Vapor Generation Methods for Chemical Warfare Agents" (U.S. Army Edgewood Chemical and Biological Center, Maryland [ECBC], Technical Report 148, March 1998). Analytical techniques specific to manual sampling of the MINICAMS® appears in "Analytical Methodology for Quantitative Determination of O-ethyl S-(2-diisopropylaminoethyl) methylphosphonothiolate (VX)" (U.S. Army Edgewood Research, Development and Engineering Center [ERDEC], Aberdeen Proving Ground, Maryland, Technical Report 476, March 1998). These methodologies, which have been used for many years, were also successfully employed in DPP testing of various detection devices. Thus, they should serve as a guide for certification testing.

For generating vapor concentrations to test IDLH detectors, "Multipurpose Chemical Agent Vapor Generation System" (ERDEC, Technical Report 424, July 1997) provides details on techniques for laboratory generation of agent vapor, including methods for generation of agent plus interference vapors. The multistaged system that is required for generating a desired vapor concentration level is described in the ECBC's Technical Report 148 for testing AEL detection devices.

The preferred method for analyzing vapor concentration is described in ERDEC's report 476 (March 1998). This method uses manual sample collection such that the sample enters directly into the preconcentrator tube of the MINICAMS. This avoids sample loss through the sampling line to achieve the most accurate measurement. The collected sample is then thermally desorbed into the GC column for separation and quantification using appropriate peak detectors.

Other thermal desorption-GC devices could be used in a similar fashion. Systems like the Agilent Dynatherm-GC-FPD/MS® will also produce similar results. This type of instrument can be used when a MINICAMS is not available. Analysis times, however, will be substantially longer than required by the described MINICAMS method.

The system to be used must be calibrated using standards made from materials of known concentration, such as the Chemical Agent Standard Analytical Reference Materiel (CASARM), if available. The calibration curve for the instrument is developed through varying the amount of the substance injected. GC output from the vapor sample collected is then correlated with the calibration curve to derive the exact vapor concentration by dividing the amount determined over the volume (sampling rate × time) of the sample collected.

The thermal desorption technique has been used with success. It is probably the only technique that is capable of analyzing the extremely low AEL concentrations within a reasonably short time. However, the thermal desorption–GC method does have limitations for analyzing high-concentration vapors. Other techniques, such as sample collection through bubbling through liquid solvent with subsequent injection into the GC instrument will be required for analyzing concentration levels that exceed the capability of thermal desorption techniques. Capability is exceeded when a reasonable sample rate and time cannot prevent the result from overwhelming the GC instrument. Under these circumstances, the sample can be collected in a solvent through bubbling. The resulting solution can then be serially diluted into the range needed for the GC analysis. Vapor transport efficiency must be properly addressed when collecting sample using all types of techniques. The preferred method for maximum accuracy is direct sampling into the collection media without use of an external transfer line.

Regardless of the sampling methodology used, techniques in sample collection are important to ensure that the sample is collected properly. For example, sample collection through bubbling solvent to analyze a low concentration level requiring an extra long sampling time is not advisable. Inefficient sample collection and solvent loss can cause large errors.

3.6.8 Detector Testing

Depending on the type of detection device, different evaluation techniques are required to provide appropriate characterization. If the item has different modes of operations, each of them needs to be addressed separately. Instruments that have alarm features are tested for minimum detectable level (MDL) using the alarm time within 2 min as a criterion. Those that do not have alarm features would also use the criterion for obtaining the maximum response signal at a certain time within the 2-min exposure. A more specific evaluation procedure for each type of detector may be required.

3.6.9 Return of Materials Exposed to CWAs

According to Department of the Army pamphlet 385-61, "Toxic Chemical Agent Safety Standards," instruments that have been exposed to CWA materials must remain under control of the government. KYO initiated a waiver request through proper channels to permit the return of materials subjected to CWA exposure to the company that developed them. The next section presents a sample risk assessment that permitted developers to retake control over materials that had been subjected to CWA testing.

3.7 SAFETY RISK ASSESSMENT FOR RELEASE OF TESTED, CONTRACTOR-OWNED MATERIALS TO CONTRACTOR

The purpose of securing a risk assessment is to allow for contractor-owned items to be tested against chemical agent vapor in an Army test facility at the U.S. Army

Edgewood Research, Development and Engineering Center (ERDEC), and to be able to return the items to contractor control. Army Regulation 385-61, Chapter 5, 5-1e(4)(a) states:

> Items decontaminated to 3X level may not be released from Government control. Some items may be released from Government control if all Federal, State, and local provisions have been met and approval is granted by the MACOM Commander.

However, there exist Command memoranda that delegate the decision authority, based on the level of risk, from the Materiel Command (MACOM) to the Chemical Biological Defense Command (CBDCOM) commander, to the ERDEC technical director for low-risk hazards.

Traditionally, contractors submitted detectors of various configurations via contract with the government, and the government would own the prototype items. Once agent testing was complete, the items would be designated as 3X and remain under government control. Given private sector interest in chemical agent detection, and the thrust of test service agreements under which government test facilities are used by private companies, companies need to conduct prototype testing that may not be associated with government agencies. For example, under the Domestic Preparedness Detector Testing Program, contractors provide their detectors for testing, and upon test completion, remove their detectors from the government site. In this case, there would be no government control (i.e., ownership) as required for chemical-agent-exposed 3X items per AR 385-61.

This author, KYO, requested the ERDEC Safety Office to develop an avenue for risk assessment to address hazards associated with unrestricted release of detectors to be exposed to chemical agents under SOP #CR9-ISP022-97C. This SOP details how the detectors are exposed to the laboratory-generated agent vapors.

The ERDEC Safety Office has since performed risk assessments on various detectors brought to government facilities for testing. Such procedures, when followed, permit the return of contaminated items to manufacturers. In the following, assessments of two detectors, the MiniRae and the M43A1® upgrade, are used as examples of how risks were assessed. Tables 3.16 and 3.17 define hazard severity and probability levels used as the criteria for assessing risk.

Table 3.16 Hazard Severity

Description	Category	Definition
Catastrophic	I	Death, system loss, or severe environmental damage
Critical	II	Severe injury, severe occupational illness, or major system or environmental damage
Marginal	III	Minor injury, minor occupational illness, or minor system or environmental damage
Negligible	IV	Less than minor injury, occupational illness, or less than minor system or environmental damage

Source: U.S. Department of the Army. Army Regulation 385-61.

Table 3.17 Hazard Probability

Description	Category	Definition
Frequent	A	Likely to occur frequently
Probable	B	Will occur several times in lifetime
Occasional	C	Likely to occur during lifetime
Remote	D	Unlikely but possible to occur during lifetime
Improbable	E	So unlikely that it can be assumed hazard will not be experienced

Source: U.S. Department of the Army. Army Regulation 385-61.

3.7.1 Assessment Scenario

All detectors are subjected to vapor challenge with HD, VX, and at least one G-agent (GD, GB, or GA). The maximum detector sampling rate is 2.0 l/min. Trials for each agent will be conducted at 25°C with low, medium, and high humidity, as well as at high temperature (52°C), low temperature (–10°C to –30°C), and room temperature (25°C) with ambient humidity. There are three repetitive trials per detector for a total of 54 exposures (27 at low concentrations and 27 at higher concentrations) for each agent. The maximum exposure for each trial is 15 sec. The sequence of test temperatures is from low (first) to high (last). After testing, each detector will be operated in a clean hood for 24 hours and then monitored.

The following assumptions are made:

- All tube structures are in proper working condition, that is, there are no cracks, broken parts, or defects, such that the only components to come into contact with agent vapors are within the tubing assembly.
- All parts that come in contact with agent vapors and cannot be decontaminated, such as filters, sieve packs, and so on, are removed and disposed of accordingly after testing per the operational SOP.
- There are no new porous or highly absorptive or adsorptive materials being introduced into the detector that can come into contact with the agent in the airstream.
- No detector has been exposed to chemical agents prior to testing.

Internal components that contact the vapor stream in the MiniRae:

- 6 in. of Teflon® tubing in the inlet probe
- 1 in. of stainless steel connector
- Disposable filter consisting of a Teflon membrane supported by an aluminum ring
- 0.5-in.-thick Teflon sensor containing two thin stainless steel grids
- 3 in. of Tygon® tubing
- Hard plastic pump with ethylene propylene diene monomer (EPDM) diaphragm
- 3 in. of Tygon tubing
- 0.5 in. exit port of acrylonitrile-butadiene-styrene type (ABS) plastic

Internal components that contact the vapor stream in the M43A1 upgrade:

- Sieve pack assembly
- Cell module assembly
- Manifold assembly
- Inlet housing assembly
- Two pump assemblies
- Charcoal filter assembly
- Silicone tubing, gaskets, seals

Low-temperature testing presents the greatest likelihood of agent deposition on the internal tubing assembly. However, because high-temperature testing will be performed last and each detector will be run for 24 hr in a clean hood to flush the system, any agent deposition on the inside surface of the tubes will likely evaporate. Any residual hazards will most likely be from sorption of agent vapor or liquid agent deposited through the wall of the tube materials. Thus, during any subsequent disassembly, there is a chance of off-gassing chemical agents posing a potential hazard. Studies have shown that negligible sorption of chemical agents occurs in Teflon and hard plastic. In addition, tests carried out at Arthur D. Little laboratory suggest that plasticization by agents is unlikely, especially in low agent concentrations. The only source of agent sorption would be from the silicone and Tygon tubing.

In the MiniRae detector, the Tygon tube dimensions are 3/16 in. × 6 in., or 22.8 cm^2. Data from a previous study showed that for concentrations less than 100 µg/l approximately 60% of agents were sorbed in 3 min over the tube's 64.52 cm^2 surface area. Even though the experimental number was recorded only at 3 min, and the amount of sorption is probably not a linear function of time, initially the sorption would be slow until a steady state is achieved. The rate will then decrease when approaching saturation state. For safety purposes and to simplify the analysis, an average rate will be used even though the exposure duration per trial is relatively short (15 sec). By interpolation, the sorption rate for each agent at the "challenge concentration" is calculated as:

$$\text{Sorption rate } (\mu g/\text{min-cm}^2) = (0.6 \times (\text{amount of agent } (\mu g) \text{ in 1 liter})) / (64.52 \text{ cm}^2 \times 3 \text{ min}) \quad (3.2)$$

Results are tabulated in column three of Table 3.18.

Using the maximum sampling rate at 2.0 l/min and the maximum exposure time is 15 sec, the amount of sorption per trial is calculated by multiplying the sorption rate by the total inner surface area of the tube exposed (22.8 cm^2) and the exposure time (0.25 min). Results are shown in column four of Table 3.18.

After each test cycle, the detector is run on the average of 3 min at room temperature to clear out residues and reset for the next trial. Previous studies show that sweeping a contaminated surface with air will remove chemical agent deposited on the surface and/or dissolved in the matrix of the material. Therefore, it is reasonable to assume that some agent off-gasses from the tube. Assuming 80% retention

Table 3.18 Calculated Results for MiniRae Detector

Agent	Challenge Concentration (μg/l)	Sorption Rate (μg/min/cm²)	Amount of Sorption per Trial, *N* (μg)	$\sum_{i=1}^{27} N(8^i)$ (μg)	Total Sorption (μg)
G	0.1 (low)	0.00031	0.00177	0.00706	0.0777
G	1 (high)	0.0031	0.0177	0.0706	
HD	2 (low)	0.0062	0.0353	0.141	3.671
HD	50 (high)	0.155	0.884	3.53	
VX	0.04 (low)	0.000124	0.000707	0.00282	0.0734
VX	1 (high)	0.0031	0.0177	0.0706	

Source: U.S. Department of the Army. Safety Risk Assessment for Release of Tested, Contractor-Owned Material to Contractor, 1997.

Table 3.19 Calculated Results for M43A1 Detector

Agent	Challenge Concentration (μg/l)	Sorption Rate (μg/min/cm²)	Amount of Sorption per Trial, *N* (μg)	$\sum_{i=1}^{27} N(8^i)$ (μg)	Total Sorption (μg)
G	0.1 (low)	0.000233	0.00177	0.0071	0.0781
G	1 (high)	0.00233	0.0177	0.0710	
HD	2 (low)	0.00465	0.0354	0.141	3.66
HD	50 (high)	0.116	0.880	3.52	
VX	0.04 (low)	0.000093	0.00071	0.00282	0.038
VX	1 (high)	0.00233	0.0177	0.0710	

Source: U.S. Department of the Army. Safety Risk Assessment for Release of Tested, Contractor-Owned Material to Contractor, 1997.

of the sorbed agent, the total amount remaining in the Tygon is $\sum_{i=1}^{27} N(.8^i)$ for each test concentration level, where N is the amount being sorbed after each trial. Results are showed in column 5 of Table 3.18.

In the M43A1 detector, the silicon tube dimensions are 3/16 in. × 4 in. or 15.2 cm². Using the same analysis as for the MiniRae detector, the sorption rate for the silicone tube can be calculated based on a previous study, which showed 90% retention of agents for silicone (Table 3.19). The calculated dosages from exposures are listed in Table 3.20.

For the MiniRae detector, in worst-case scenarios, 0.0777 μg of G-agent, 3.68 μg of HD, and 0.0734 μg of VX can be retained in the tube. Respective concentrations would be 0.0000777 mg/m³ for G, 0.00368 mg/m³ for HD, and 0.0000734 mg/m³ for VX. These concentrations for HD and VX exceed AELs (in Army Regulation 385 61) by 20% and 700%, respectively. However, this assumes that all agents will be off-gassed into the air at once, whereas in the real world, such off-gassing is not rapid. In addition, as mentioned above, each detector is run in a clean hood for at least 24 hr prior to monitoring, thereby allowing even more off-gassing

Table 3.20 Dosage from Exposures

Agent	Challenge Concentration (mg/m³)	Total Amount Passing Through Each Detector (mg)	AEL Value (mg/m³)
VX	0.04 to 1.0	0.014	0.00001
GD	0.10 to 1.0	0.0149	0.00003
GB			0.0001
GA			0.0001
HD	2.0 to 50.0	0.702	0.003

AEL, airborne exposure level.

Source: U.S. Department of the Army. Safety Risk Assessment for Release of Tested, Contractor-Owned Material to Contractor, 1997.

to take place, lowering the potential amount of agent present. Finally, the SOP requires that monitoring be performed to ensure that the items are at the 3X level, and as contamination free as possible. Taking these considerations into account, the residual hazard remaining would be from VX off-gassing due to the high percentage ratio of amount sorbed to the AEL. Although the 0.0734 μg of VX does not present an appreciable liquid contact hazard, the tubing should still be considered potentially contaminated and should only be used under engineering controls. It is recommended that Tygon tubing in particular be disposed of and replaced prior to release of detector to the contractor.

For VX we can compare the maximum possible concentration given above as 0.0000734 mg/m^3 vs. a physical effects level of slightly less than 0.09 mg-min/m^3, where an effect like mycosis might become noticeable. Even when this level is dropped an order of magnitude, it would require 0.009 mg-min/m^3 divided by 0.0000734 mg/m^3, which is equivalent to 123 min of exposure for significant (e.g., noticeable effect that could be detected by the operator) exposure to occur with VX.

In the case of HD, using the maximum 2 mg-min/m^3 safe air concentration for the eyes, based on the MSDS, more than 2 mg-min/m^3 divided by 0.036 mg/m^3 (equivalent to 55 min of exposure) are necessary to pose a serious risk to the eyes.

However, the above concentrations could not be maintained for any significant length of time, since they rely on the concentration of both VX and HD to remain constant. Realistically, even in a worst case scenario, where there is no ventilation:

- Actual agent concentrations would be lower (e.g., off-gassing would proceed over time).
- There would be some movement of air about the room dispersing the agent.
- The agent would diffuse into the entire room, rather than remain confined at the concentrations mentioned above in the area where the worker is exposed.

Thus, even without the addition of engineering controls, the severity of the hazard is categorized as III, marginal.

Using similar calculations for chemical sorption of silicon tubing in the M43A1 detector in worst-case scenarios, 0.0781 μg of G, 3.66 μg of HD, and 0.0380 μg of VX could be sorbed in the tube. Again, using the same scenario as above with the MiniRae detector, if these amounts of agent are off-gassing into 1 m^3 of air, the concentration would be 0.0000781 mg/m^3 for G, 0.00366 mg/m^3 for HD, and 0.0000380 mg/m^3 for VX. These concentration values are approximately the same as that of the MiniRae in the analysis. Therefore, severity and suggested recommendations would also be the same, and severity is also categorized as III, marginal.

If recommendations given below are implemented, particularly the 24-hr purge, monitoring to 3X levels, removal of porous parts exposed to the agent, and replacement of tubing, the probability of agent exposure for contractor personnel is categorized as D, "remote probability."

The greatest hazard risk is potential exposure of workers when they disassemble the detector and potentially release VX vapor agents. There is also a concern with respect to allowing any exposure to HD, no matter how minor, since it is a carcinogen. Since the VX value is higher than the AEL, and we want to minimize the possibility of any exposure to HD, the following recommendations should be provided to, and agreed on, by the contractors:

- Any initial disassembly operation will be done under a properly ventilated chemical fume hood.
- Workers will wear personal protective clothing and equipment (gloves, apron), and have an appropriate mask approved by the National Institute for Occupational Safety and Health (NIOSH) readily available.
- If there are signs of liquid depositions in the detector, those areas should be decontaminated with a 5% chlorine bleach solution immediately.

These actions should reduce the probability of exposing contractor personnel to any possible agent vapor even if there is an internal pathway leak to the electronic components because the hazard only consists of a single release when the unit is first opened.

In addition, the government and contractor effectively agree to accept a particular risk level before tested items are returned. ERDEC Engineering Directorate personnel agree to carry out the following activities on the contractor's detector after testing:

- Run the detector test from low to high temperatures to minimize residual agent condensing within the detector
- Ensure that any Tygon or silicone tubing, filters, or any porous materials within the air sampling path are removed and disposed of accordingly
- Decontaminate all detector surfaces
- Run each detector in a clean, agent-free chemical hood, for at least 24 hr after the last exposure
- Monitor the detector to the 3X level to ensure that there is no off-gassing of residual agents

When these activities are implemented, the risk to workers is categorized as low, at severity III and probability D.

Statement for Releasing of XXX Monitored Items to Contractor: "We, (contractor name), understand and acknowledge that while every precaution has been made to minimize the potential for any residual chemical agent material to be present internally to the detectors provided for testing, that there is a remote chance chemical agent may be present. We, (contractor name), agree to take appropriate precautions when allowing personnel to deal with these items, to include incorporating initial disassembly conducted under engineering controls and with appropriate personnel protection to preclude any exposure. We, (contractor name), accept responsibility to ensure that personnel allowed access to these items are informed of previous testing until such a point that we have determined there is no residual hazard associated with these items. We, (contractor name), will also ensure that any entity, government or private, which we allow access to or which agree to purchase these detectors, are told of the potential risk associated with these items."

CHAPTER 4

Vapor Generation Techniques

CWAs are manufactured and stored by government facilities for basically two reasons. Stored CWAs serve as an effective deterrent to "first use" of CWAs by others. Secondly, manufacture of these agents permits necessary research of agent characteristics, and provides a means for research of detection techniques and antidote developments. Early detection and warning of even minute amounts of CWAs and/or TICs in the air are essential to minimize casualties.

To develop effective detection devices, methods to evaluate their effectiveness are prerequisites. Open-air dissemination of toxic compounds for testing purposes has been banned in the United States for decades. Researchers resort to diverse methodologies and techniques to create test atmospheres/environments. Much effort has been focused on simulating real-world scenarios in the laboratory.

As mentioned in Chapter 3, sampling is one of the most difficult steps in testing the capabilities of chemical and biological agent detectors. Methods to produce and deliver a controlled stream of air with a precisely controlled concentration of the target compound are crucial for evaluating detector performance. Problems arise in ensuring that the sample presented to the detector during the evaluation represents a realistic scenario. Also, physical property measurement studies, such as vapor pressure, adsorption equilibrium, and toxicity are required to assess contamination spread to derive physical hazard assessments. All of these studies rely on precisely generated streams of air in which the concentration of target chemicals is stable. In this chapter, we discuss common airstream generation techniques used by researchers.

Safety precautions are the utmost importance when working with highly toxic chemicals. Mishandling can cause harm not only to laboratory workers but also others in the vicinity. Chemical behavior and all safe-handling procedures must be understood before attempting to generate these toxic vapors. The material safety data sheet (MSDS) (see Appendix A for example) must be understood, and a specific standing operation procedure (SOP) should be developed and followed to minimize the risk of accidental exposure to the toxic components. The SOP should detail all necessary precautions and operational instructions. Potential hazards are spelled out. The SOP is especially critical when the application involves the use of lethal chemical agents or other highly toxic materials. A single mistake could cause catastrophic fallout, including injury or death to operators and others.

The SOP is generally prepared by the principal investigator who lists all conceivable precautions throughout the process from beginning to the end. The organization safety office must review and assess the risks. A dry run (operation minus the toxic material) through the steps is performed before approval is granted. During the dry run, the procedures are observed by a panel of knowledgeable individuals who provide comments and critiques.

Before an operation begins, operators must read and sign the SOP indicating that all steps have been understood. Pre- and postoperation checklists ensure that the apparatus, protective equipment, engineering controls, and other items involved are in proper order. The checklists will also serve as a reminder to the operator of precautions.

Operations involving the use of CWAs are considered as surety operations. All surety operations require a buddy system, that is, two people who are knowledgeable and qualified for working with surety materials must be present before and during the operation. These individuals must undergo an extensive clearance process called the Chemical Personnel Reliability Program (CPRP) to guarantee their integrity and proper training for working with controlled materials. In the event of mishap, a buddy would presumably be available to rescue the other person or secure help.

An abbreviated SOP sample appears in Appendix B. It shows the necessary contents of a typical SOP. Safe handling of toxic materials cannot be overemphasized.

4.1 GAS LAW AND GAS CONCENTRATION

To fully understand the reasons behind vapor generation methods, reference to a few basic laws of physics is necessary. Physical, chemical, and gas laws dictate the behavior of a gas under different conditions. For purposes of simplicity, under normal circumstances, a gas generated in the laboratory is assumed to behave similarly to the ideal gas. We briefly review the Ideal Gas Law for its applicability to various generation techniques, and provide a method for vapor concentration calculation.

4.1.1 Mole, Molar Weight, Molar Volume, and Mole Number

Three important concepts used in the following discussion are mole, molar weight, and mole number. A "mole" of a given substance contains 6.02×10^{23} molecules, commonly known as Avogadro's number. For example, 1 mole of nitrogen gas contains 6.02×10^{23} nitrogen molecules. Similarly, a mole of water contains 6.02×10^{23} water (H_2O) molecules.

Molar weight is the weight of 1 mole of a substance that is equivalent to the atomic weight, molecular weight, or formula weight in grams. For example, the molecular weight of ammonia is 17 atomic mass units (amu). Therefore, its molar weight is 17 g, which contains 6.02×10^{23} molecules. According to the Ideal Gas Law, 1 mole of a gas, at 0°C and standard sea-level atmospheric pressure of 760 mm of mercury (Hg), occupies 22.4 liters of space.

The mole number is the number of moles that a specific amount of a substance represents. For a gas, it can be calculated based on its weight or the volume it occupies. For example, 34 g of ammonia is equal to 2 moles (34 g/17 g) of ammonia; therefore, its mole number is 2. At 0°C and standard sea-level atmospheric pressure of 760 mm of mercury (Hg), these 34 g of ammonia with a mole number of 2 would occupy a volume of (2 × 22.4) 44.8 liters.

4.1.2 Ideal Gas Law

The Ideal Gas Law, also known as Boyle's Law, characterizes the behavior of an ideal gas. An ideal gas is a hypothetical gas with negligible molecular size and intermolecular forces. Boyle's Law states that for an ideal gas, the relationship among its pressure (P), volume (V), and absolute temperature (T in Kelvin) can be expressed as:

$$PV = nRT \tag{4.1}$$

where n is the number of moles of the gas, and R is the molar gas constant (8.3143 joules per Kelvin per mole). For n moles of an ideal gas, Equation 4.1 can be rearranged as follows:

$$nR = \frac{PV}{T} \tag{4.2}$$

Equation 4.2 states that for a certain amount (n moles) of an ideal gas, the relationship among P, V, and T remains constant under various temperature and pressure conditions. Therefore,

$$nR = \frac{P_1V_1}{T_1} = \frac{P_2V_2}{T_2} \tag{4.3}$$

Equation 4.3 can be expressed simply as

$$\frac{P_1V_1}{T_1} = \frac{P_2V_2}{T_2} \tag{4.4}$$

Example 4.1

As stated above, 1 mole of an ideal gas at 0°C and 1 atm (760 mmHg) will occupy a volume of 22.4 l. At room temperature (25°C, or 298 K) and 1 atm, the volume

(V_1 in the following equation) for 1 mole of ideal gas can be calculated using Equation (4.4) as 24.45 l:

$$V_1 = \frac{T_1}{P_1} \times \frac{P_2 V_2}{T_2} = \frac{298(K)}{760(mmHg)} \times \frac{760(mmHg) * 22.4(l)}{273(K)} = 24.45(l)$$

The volume of a vapor is directly dependent on temperature, but is inversely dependent on external pressure. In other words, a gas will expand and occupy a larger volume as the temperature is increased or when the external pressure is reduced.

4.1.3 Vapor Concentration

The concentration of a vapor is generally expressed as either volume concentration of parts per billion (ppb) or parts per million (ppm), or as weight concentration such as milligrams per cubic meter. At concentrations higher than 0.1% (>0.1% v/v), volume percentage is normally used instead of ppb or ppm. In this book, except those specially noted, volume concentrations are expressed as ppm, ppb, or percentage (%) concentration. For example, the concentration of ammonia in a gas mixture containing 1 μl (microliter) of ammonia in 1 cubic meter ($1\ m^3 = 10^9\ \mu l$) of the gas mixture is $1\ \mu l/10^9\ \mu l$ or 1 ppb. If the gas mixture contains 1 ml of ammonia, the concentration is 1 ppm. One liter of ammonia gas in 1-m^3 gas mixture will produce a concentration of 1000 ppm that is equivalent to a 0.1% concentration.

Concentration can also be expressed as weight-per-volume (w/v) concentration, such as milligrams per cubic meter. The weight-per-volume concentration expresses the actual amount of the substance contained in a fixed volume. The expression is applicable for any pressure and temperature condition. The weight-per-volume concentration and volume concentration can be converted by applying the gas law and mole volume constant explained earlier. Appearing below is an example that illustrates the relationship between volume concentration and weight concentration.

Example 4.2

Assume 1 liter of a 1% ammonia gas mixture. What are the equivalent weight concentrations (in milligrams per cubic meter) at 0°C and 25°C? First, we determine the equivalent volume of ammonia in a liter of the 1% ammonia mixture that is equivalent to 0.01 l. The molar volumes at 0°C and 25°C are 22.4 l and 24.45 l, respectively, as discussed earlier. The mole numbers of ammonia in this gas mixture at the two temperatures can be calculated as follows:

$$n = \frac{V}{MV}$$

where MV is the molar volume at 0°C (22.4 l) and 25°C (24.45 l).

Table 4.1 Calculated Equivalents of 1 Liter of 1% NH_3 (Results for Example 4.2)

Temperature (°C)	Mole Number	Weight (g)	Concentration (mg/l)	Concentration (mg/m³)
0	0.000446	0.007589	7.589	7589
25	0.000409	0.006953	6.953	6953

Given that the molar weight of ammonia is 17 g, the weight of ammonia under the two conditions is calculated as

$$W = n \times MW = 17n(g)$$

By converting the weight to milligrams and the volume to cubic meters, the weight concentrations in milligrams per cubic meter are obtained. Results are listed in Table 4.1.

The above example shows that even though the stated volume concentrations are the same at different temperatures, the amounts of substance (weight) contained in fixed-volume concentration gas mixtures are different.

Similarly, weight concentration (milligrams per cubic meter) can be converted to volume concentration. To do this, the above calculation is reversed. The weight is converted into a molar number first, and then, based on the given temperature and pressure, its volume and volume concentration are calculated. Table 4.2 lists the conversion factors for common CWAs and high-hazard TICs. The factors can be used to easily calculate the weight-volume concentration from volume concentration. For example, suppose that the concentration of generated GA vapor is 0.02 ppm at 25°C, the weight-volume concentration is calculated at equal to 1.33 mg/m³ by knowing that 1 ppm of GA at 25°C is equal to 6.63 mg/m³:

$$C = 0.02(ppm) \times 6.63(\frac{mg/m^3}{ppm}) = 1.33(mg/m^3)$$

4.2 VAPOR GENERATION

A chemical vapor generation system consists of the vapor source and volume dilution through a mixing device. The controlled vapor from the vapor source is introduced into the dilution flow and mixing device to achieve the desired concentration. The process is very simple. Precise control of various flow blends and understanding pertinent basic physical chemical laws determine the suitability of a given dissemination technique.

Several vapor generation methods use direct contact of the liquid with a carrier gas stream. These include sparging (the carrier gas bubbles through the liquid); syringe injection of the liquid material directly into a carrier gas-flow stream; and head-space vapor saturation. Solid-state vapor generation exploits the use of vapors

Table 4.2 Conversion Factors for CWAs and Selected TICs at Three Temperatures

Chemicals	MW (amu)	Conversion Factor at Selected Temperature ([mg/m³]/ppm) 0°C	25°C	40°C
GA	162.13	7.24	6.63	6.31
GB	140.10	6.25	5.73	5.46
GD	182.12	8.13	7.45	7.09
GF	180.2	8.04	7.37	7.02
VX	267.38	11.94	10.94	10.41
HD	159.08	7.10	6.51	6.19
HN-1	170.08	7.59	6.96	6.62
HN-2	156.07	6.97	6.38	6.08
HN-3	204.54	9.13	8.37	7.96
L	207.35	9.26	8.48	8.07
AC	27.03	1.21	1.11	1.05
CK	61.48	2.74	2.51	2.39
CG	98.92	4.42	4.05	3.85
DP	197.85	8.83	8.09	7.70
Ammonia	17	0.76	0.70	0.66
Arsine	77.93	3.48	3.19	3.03
Boron trichloride	117.2	5.23	4.79	4.56
Boron trifluoride	67.8	3.03	2.77	2.64
Carbon disulfide	76.1	3.40	3.11	2.96
Chlorine	70.91	3.17	2.90	2.76
Diborane	27.7	1.24	1.13	1.08
Ethylene oxide	44.1	1.97	1.80	1.72
Fluorine	38.0	1.70	1.55	1.48
Formaldehyde	30	1.34	1.23	1.17
Hydrogen bromide	80.9	3.61	3.31	3.15
Hydrogen chloride	36.5	1.63	1.49	1.42
Hydrogen fluoride	20.0	0.89	0.82	0.78
Hydrogen sulfur	34.1	1.52	1.39	1.33
Nitric acid (fuming)	63.0	2.81	2.58	2.45
Phosphorus trichloride	137.4	6.13	5.62	5.35
Sulfur dioxide	64.1	2.86	2.62	2.50
Sulfuric acid	98.1	4.38	4.01	3.82
Tungsten hexafluoride	297.9	13.30	12.18	11.60

Note: Concentration in milligrams per cubic meter = concentration in parts per million × conversion factor.

preadsorbed onto high surface-area solids below saturation capacity in order to greatly reduce the equilibrium vapor pressure of liquids and thus permit direct generation of vapors at lower flows and concentrations. Similarly, liquid materials can be diluted, using a noninterfering solvent, to reduce substance's volatility to achieve similar results. Other "indirect" methods, where a barrier is introduced between the liquid and vapor such as permeation, diffusion, and effusion can also be employed. For chemicals with very low volatility, vapor generation may not produce the concentrations required to effectively evaluate a given detection device. Aerosol generation is an alternative technique. Aerosols can be generated using nebulizer spray or vapor condensation techniques. For some gaseous compounds, chemical reactions may be necessary if analytically prepared cylinders are not commercially available.

Agent concentration generated by various methods must be analyzed to verify and determine the reliability of each technique. Theoretically calculated concentrations can only be used as estimates of the generation process. They cannot substitute for actual concentrations because of the adsorption and absorption effects by generator plumbing and the efficiency to deliver the saturated vapor before dilution. Sample collection techniques and some pertinent analytical methodologies will be discussed. It should be noted that many analytical methodologies are currently used by laboratories and researchers for diverse chemicals. In this chapter we present techniques that have been successfully used for many years.

4.2.1 Vapor Generation Methods

Several methods to generate controlled vapor concentrations that are currently in use are briefly described in this section. Each vapor generation technique serves a specific purpose. The most important difference among these techniques is the range of the vapor concentration generated.

4.2.1.1 Evaporation Method

The most direct method of vapor generation is to pass all or part of a carrier stream through a liquid reservoir to produce the vapor concentration. Since the chemical continually evaporates into the carrier gas stream, vapor containing the chemical is constantly generated for the test. The vapor concentration in the resulting carrier stream depends on the vapor pressure of the liquid, and the efficiency of the carrier stream used for the sparger generator. This method has been used extensively to generate airstreams containing CWAs and simulants to characterize the destructive efficiency of low space-time catalytic oxidation reactors.

The evaporation method used for a liquid or solid chemical is based on the evaporation rate or volatility of the chemical. Concentration of the vapor generated with this method may be very high for highly volatile compounds. On the other hand, it may not produce sufficient vapor from low-volatility liquids or solids. To use this method, the suitable dilution gas, such as dry air or nitrogen, is introduced to a gas dispersion bottle containing the chemical to be tested. The gas that passes over or through the liquid will carry the substance vapor that is used as the source vapor for the vapor generation. Figure 4.1 shows an example of a type of gas dispersion bottle that is used to generate vapor with the evaporation method.

Vapor concentration depends on several factors such as the size of the dispersion bottle, the temperature of the liquid, and the carrier gas flow rate. It can be expected that the carrier gas at a higher flow rate may not be vapor saturated at the outlet since it is constantly passing through the liquid without sufficient residence time to reach liquid — vapor equilibrium. Increasing the volume above the liquid or decreasing the flow rate of the dilution gas prolongs residence time and increases the saturation level of the generated vapor at the outlet. If equilibrium can be established,

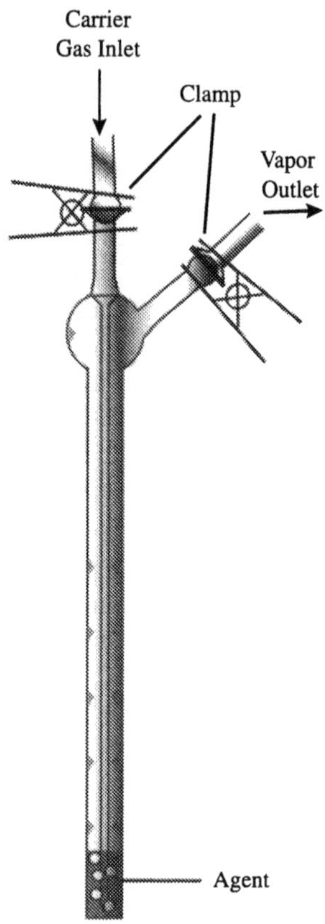

Figure 4.1 Example of gas dispersion bottle.

the concentration of the substance in the carrier gas after it passes through the liquid is estimated as follows:

$$C(\text{ppm}) = \frac{VP(\text{mmHg})}{760(\text{mmHg})} \tag{4.5}$$

where VP is the substance's vapor pressure at a certain temperature. This estimation is based on the assumption that the vapor is saturated with the target chemical. In general, this equation overestimates vapor concentration. The carrier gas may not be saturated with the chemical vapor at high carrier flows.

Many practitioners also calculate average concentration based on the gravimetric method during a test. The gravimetric method measures the weight change after a test period. The evaporation rate (R_{eva}) of the chemical is calculated as follows:

$$R_{eva} = \frac{W_0 - W_1}{t} \tag{4.6}$$

where W_0 is the weight (g) of the chemical at the beginning of the test, W_1 is the weight (g) of the chemical at the end of the test, and t (minutes) is test duration.

When the flow rate of the carrier gas is controlled as a constant during the test and temperature change is minimal, concentration can be calculated as

$$C(g/l) = \frac{W_0(g) - W_1(g)}{t(\min)} \times \frac{1}{F(l/\min)} \tag{4.7}$$

where F is the flow rate of the carrier gas (in liters per minute). This gravimetric method of determining concentrations is also applicable to permeation and diffusion methods of vapor dissemination, which are discussed later in this chapter.

Varying the temperature of the reservoir can change vapor pressure of the substance within. For example, by placing the dispersion bottle in a temperature-controlled hot-water bath (Figure 4.2), the vapor pressure of the substance is increased and, consequently, the concentration of the substance in the carrier gas is increased.

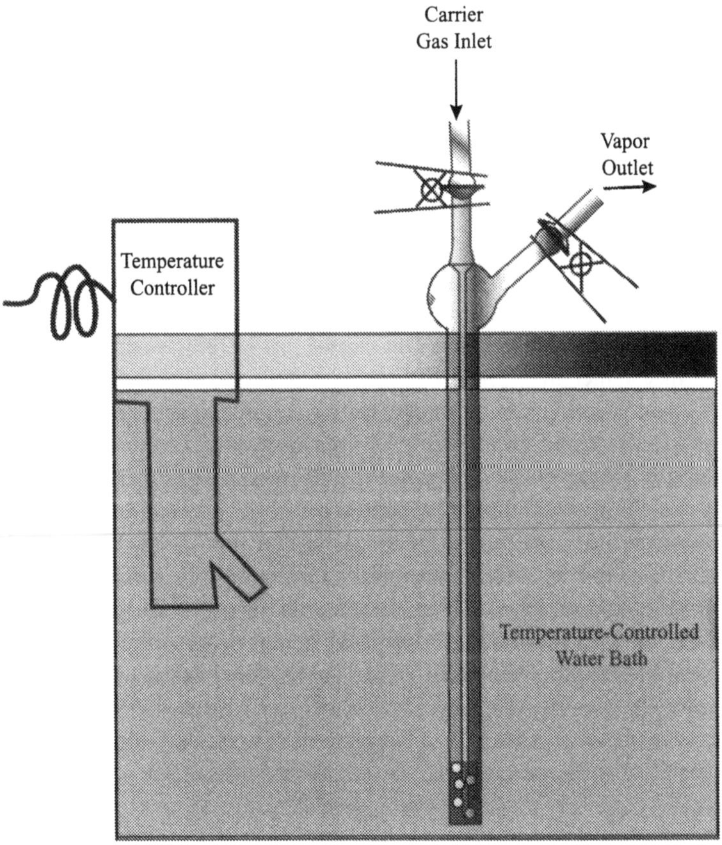

Figure 4.2 Water bath used with gas dispersion bottles for vapor pressure control.

Often a solution of high-vapor-pressure substance in a low-vapor-pressure solvent is used as a vapor source to control and lower the generated vapor concentration. The volatility of the high-vapor-pressure substance (the solute) can be reduced substantially to permit generation of lower concentrations without having to use large volumes of dilution air. Certain important issues should be considered when using this method, however. The vapor generated from the solution will contain not only the substance of interest but also the solvent. The chosen solvent must not interfere with the test. Since evaporation rates of the substance and the solvent are different, the vapor pressure of the substance continually changes with time as it is being depleted at a much higher rate than the solvent. Thus, the concentration of the solution changes during the test. Calculating the concentration based on the gravimetric method may not be appropriate.

4.2.1.2 Saturation Method

The heart of a vapor generation process is the ability to consistently carry the vapor from the substance source. Ideally, the carrier stream carries the substance vapor in its saturate form. Degree of saturation depends on relative flow rate and substance volatility. Two types of saturator will be described. First, the delta tube saturator is used for low total flow applications where a high-purity vapor stream is required. Second, a more conventional multiple-pass saturator equipped with a high surface-area wicker facilitates evaporation for greater flow demands.

4.2.1.2.1 Delta Tube Saturator

The delta tube generation system described here is primarily used for testing of small detection devices in the laboratory. The sparger or "bubbler" type of generator described above is replaced by a triangular shaped "delta tube," and a small drop (e.g., ≤100 μl) of agent is used. The carrier gas stream sweeps over the liquid or solid surface rather than through it to carry the vapor for further dilution (Figure 4.3). A two-stage dilution technique can be employed depending on the targeted concentration range and the volatility of the compound of interest.

The delta-tube generator system is the most useful generation system for point-sample detector testing. It permits generation of very low concentrations of highly purified vapor using minimal dilution flows. The use of the delta tube design has also eliminated the dangerous backsplash of toxic substance that was possible when using bubbler generator types. Because of internal pressure within the vapor generator system, malfunction or disconnection upstream from the sparger in the flow system could cause the substance in the bubbler to flow backward and thus cause backsplash. The delta tube generator system is constructed using commercially available components. Corrugated Teflon lines, Swagelok® fittings, and glass ball-and-socket joints are used to minimize sorption losses and maximize installation flexibility.

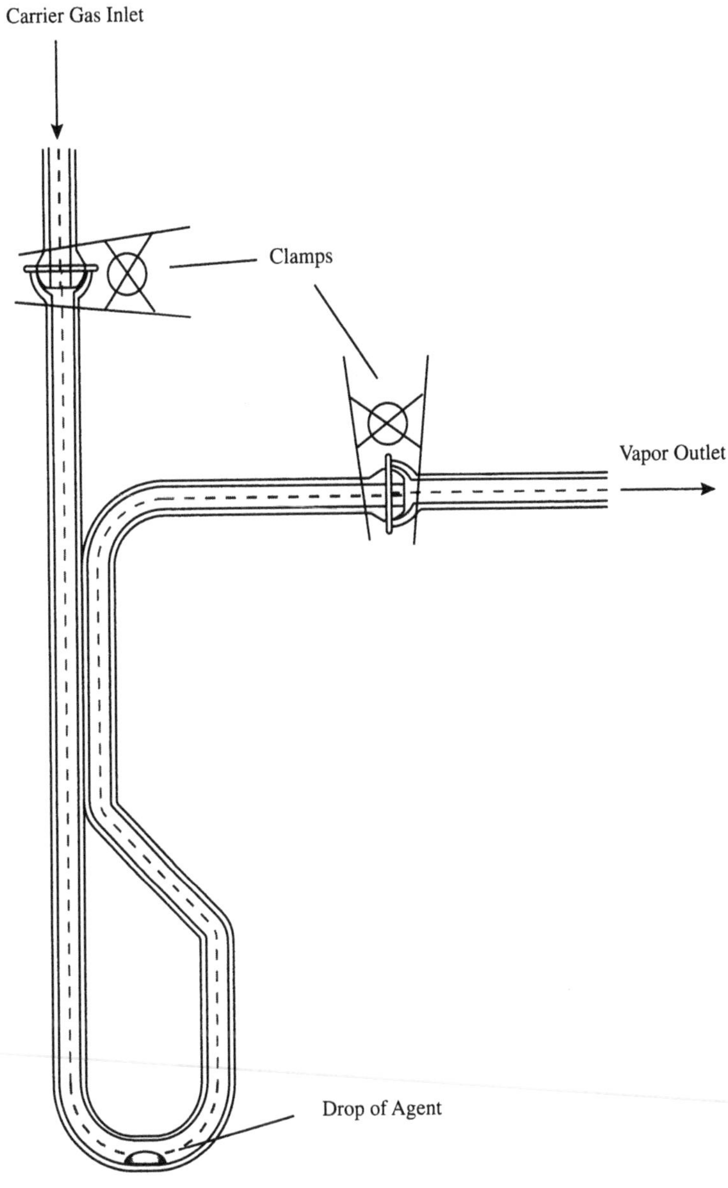

Figure 4.3 Delta tube.

4.2.1.2.2 Large-Scale Saturator

This device is similar to the delta tube used to generate vapor for various applications, including physical property (vapor pressure) measurements and filter testing. This larger version is shown schematically in Figure 4.4. Comprised of concentric glass and ceramic tubes up to 5 cm in diameter and 15 cm long, it has been used to

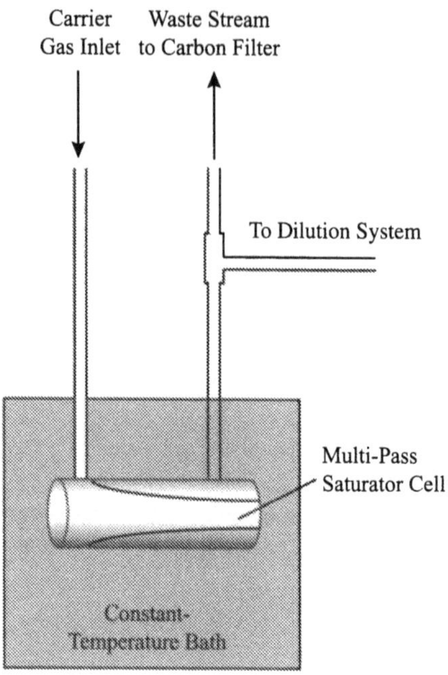

Figure 4.4 Large-scale saturator.

generate flow rates up to 10 l/min. Larger systems can be scaled down to less than 0.1 l/min, and can hold up to 100 g of liquid. These larger generators and liquid samples make it more difficult to purify low-volatility agents such as VX, however. Other than that, these saturators have proven to be extremely stable and easy to operate. Large-scale saturators have been used extensively to generate challenge streams for CWAs and simulants studies where as much as 10 g per experiment are needed to assess filtration performance.

4.2.1.3 High-Pressure Injection Method

During recent laboratory and subscale tests of a pressure-swing adsorption (PSA) filtration system (Causey and Buettner, 1996), it became necessary to introduce a controlled mass (dose) of medium-volatility vapors, including the nerve agent GB, into a high-pressure, high-flow feed stream, and to ensure evaporation prior to filter entry. The liquids are injected into a large (200-gallon) mixing chamber that is then pressurized to the feed condition (approximately 65 psi). Sufficient time is allowed for the liquid to evaporate and mix prior to injection. After the premixed sample is prepared, the experiment was initiated by feeding high-pressure air to the PSA feed stream. Once the PSA system achieved steady state at the desired feed humidity, the flow was diverted through the mixing chamber prior to entering the PSA test rig.

4.2.1.4 *Diffusion/Effusion Method*

Molecules tend to move from a high-concentration area toward a low-concentration area, and the flux is a function of the concentration gradient. Therefore, by creating and keeping a concentration gradient near the surface of the source chemical, it will be continually released into the dilution gas stream through the diffusion process.

The diffusion method can be used to deliver higher-concentration vapor than the permeation method (discussed in next section) but lower concentration than the evaporation method. It can produce vapor with concentration in the parts-per-billion to 100-ppm range and is used for generating vapor from liquid chemicals.

Figure 4.5 shows an example of a diffusion device. A small vial containing the source chemical is placed in a larger container in a temperature-controlled oven.

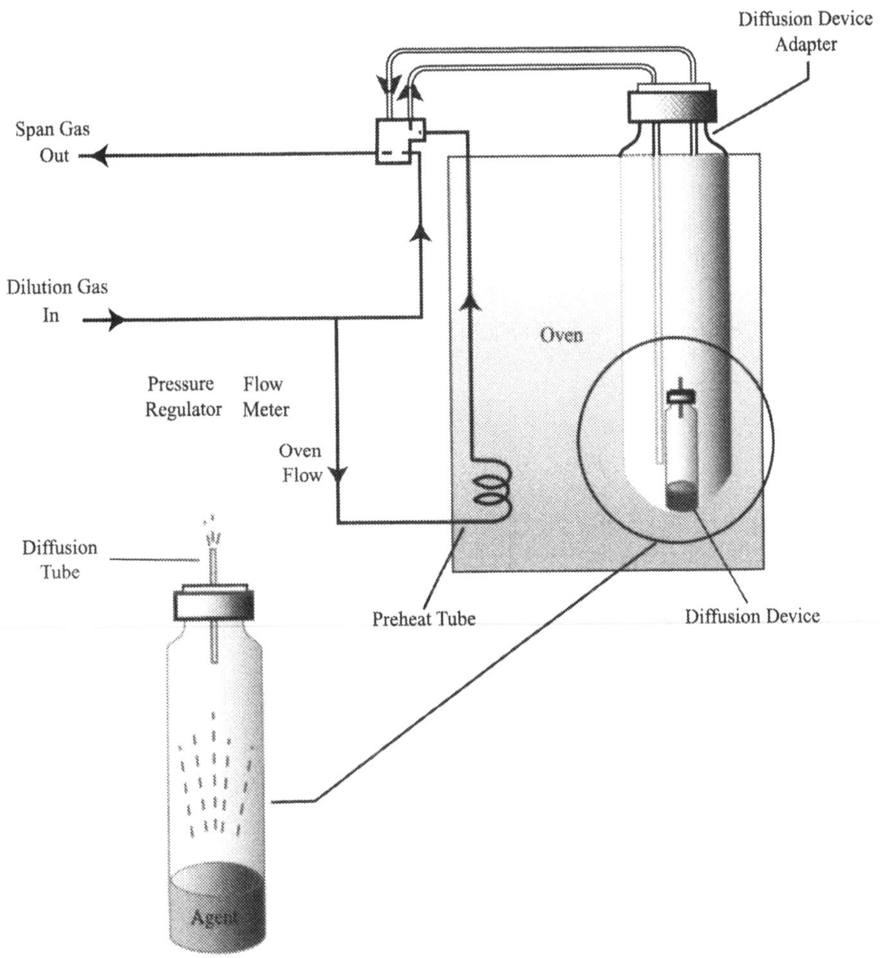

Figure 4.5 Diffusion device.

The substance in the vial diffuses out of the vial through the diffusion tube as the vapor source is swept away by the dilution gas. Changing the temperature of the oven or diameter of the diffusion tube will change the diffusion rate. Higher temperature and larger diffusion tube diameter will permit faster diffusion. Increasing the diffusion tube length, conversely, reduces diffusion flux. Manipulation of these combined effects permits controlled vapor generation.

4.2.1.5 Permeation Method

The permeation method is based on the slow penetration or permeation of molecules through a polymeric membrane. The permeation rate of the compound is determined by physical characteristics of the chemical and the membrane. Permeation is pressure and temperature dependent. Higher vapor-pressure compounds will have a higher permeation rate. Similarly, high temperature will create high internal vapor pressure that leads to higher permeation rate.

Under suitable conditions, permeation tubes can be used as the source for generating low-vapor concentrations. The mass flux of a substance through the wall of the permeation tube is constant at a given temperature, and produces a predictable concentration as a function of the carrier flow rate. Figure 4.6 is a schematic of a permeation tube generator, showing a system in which the tube temperature and flow rate are controlled in order to achieve the desired final concentration.

Controlled and predictable concentrations over a wide flow rate range are readily achievable. The generator uses a small diameter tube (e.g., 1/4 in. O.D.) made of a suitable membrane material such as polytetrafluoroethylene (pTFE) loaded with the liquid agent. The compound contacts the inside surface of the membrane and the vapor slowly permeates.

When the permeation tube is held at a constant temperature, a steady flux flow of compound vapor is emitted. This output can be measured by weight loss over a period of time to determine the permeation or emission rate. The permeation rate is generally expressed as number of nanograms per minute (ng/min). Weight-loss measurements can be taken at various temperatures to produce different emission rates for the same tube. The emission can be diluted with a known flow of dilution air to achieve the desired concentration.

Permeation tubes for many liquid chemicals, such as acetone and certain CWA simulants can be purchased commercially. However, permeation tubes for CWAs or TICs are not readily available and must be fabricated in the surety laboratory. Figure 4.7 shows a piece of Teflon tubing (e.g., 10 to 20 cm long and 1/4 in. diameter) is sealed with a Teflon plug at one end (Figure 4.7b). The chemical is then injected into the tube to fill about half of the tubing length from the other end (Figure 4.7c) and sealed with another Teflon plug (Figure 4.7d). The seal must be tight so that the chemical cannot leak from the ends.

The permeation rate is determined by the gravimetric method. The permeation device, after it has been assembled, is weighed and then placed in a temperature-controlled cavity with constant flow of nitrogen or clean air to sweep away the permeated chemical vapor. After a period of time, perhaps days, the tubing is weighed again and the permeation rate is determined by weight loss over elapsed time. Since

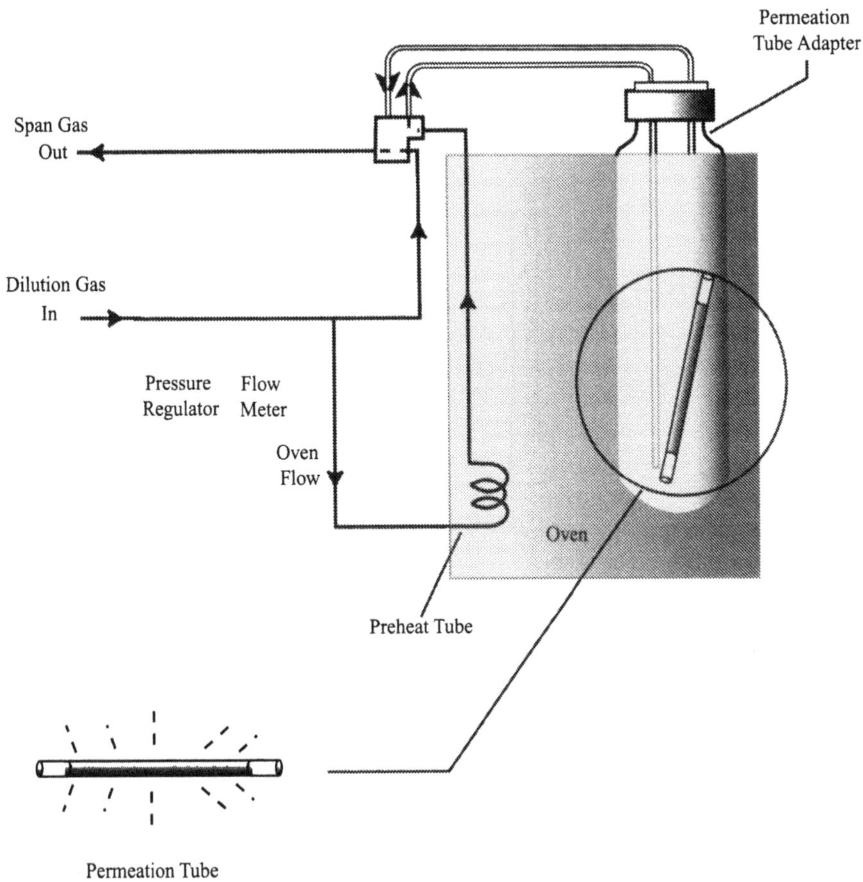

Figure 4.6 Permeation device.

the permeation rate is a constant at a constant temperature, during tests, increasing of the dilution flow rate decreases the generated vapor concentration.

4.2.1.6 Syringe-Pump Injection Method

The syringe-pump injection method has been used for a variety of applications mainly to produce higher concentration and volume generations that may not be achieved through usual volatilization techniques. The technique is accomplished by directly injecting the liquid into a carrier stream through a syringe at a fixed rate. Low-concentration streams are generated by injecting as little as 0.1 mg/min of GB into an air stream at a 1000 to 1500 l/min flow rate for a final concentration of approximately 0.1 mg/m^3.

A similar vaporization method has been developed by forcing liquid into a stream of heated air. If enough heat is applied, the liquid will be instantaneously evaporated. This method has been used to generate a high-volume stream of simulant for large-item testing in the chamber. Concentrations of 100 mg/m^3 or higher can readily be generated and maintained using this method.

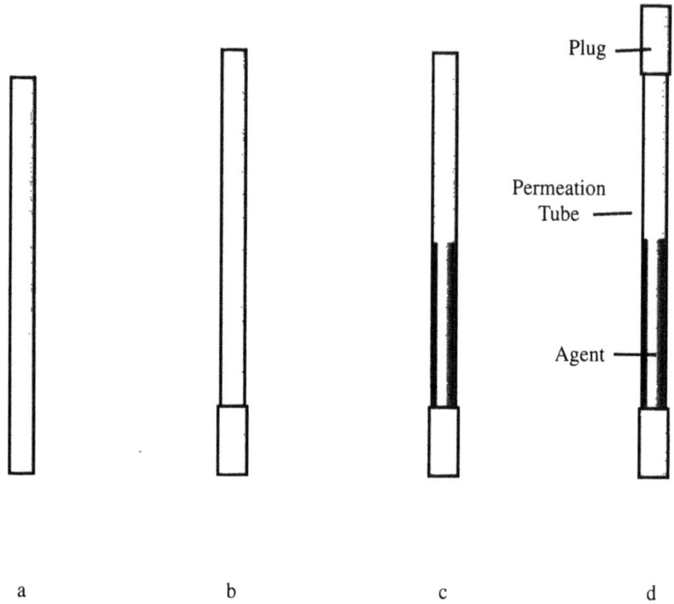

Figure 4.7 Procedures to assemble permeation tube.

This method generates vapor at a rate of 400 cubic feet per minute (cfm), which is equivalent to approximately 11.3 m^3, using an electric blower attached to the end of a 6-in.-diameter aluminum pipe. Two 208-V heater elements are mounted inside the pipe. This length of pipe is then inserted into the side of an 8-in.-diameter length of pipe roughly at a 45° angle. The 6-in. pipe is inserted into the 8-in. pipe at least 0.5 in. to prevent any liquid simulant from entering the pipe containing the heaters. The length of 8-in. pipe is mounted vertically in a stand. Then, the simulant, which is volume controlled by a needle valve, is pumped through an orifice mounted near the top of the 8-in. pipe to form a spray on its sides. The spray is vaporized by the hot air created by the heater elements and forced through the generator by the blower motor. Temperatures of 52°C (125°F) have been shown to vaporize nearly all the simulant put into the system. Since not all the simulant will vaporize, a drain must be installed in the bottom of the 8-in. pipe to recover liquid for reuse. An approximate material balance can be calculated if the airflow and simulant injection and recovery rates are measured.

Simulant vapors released into the test chamber are stirred with fans to produce a uniform challenge. The feed is monitored with proper analytical devices and data acquisition system. This system has been employed to generate a high volume of vapor in a large chamber for testing the protection efficiency of unhardened collective protection shelters. These shelters typically use filtered air to overpressurize the inside of the shelter to preclude infiltration of unwanted vapors.

A variation of this method is to inject liquids with low boiling points directly into process streams by virtue of their own head pressure. Cyanogen chloride (ClCN), for example, is frequently used to test developmental and fielded filtration materials and systems. It boils at 12.8°C. Therefore, it has an inherent high vapor pressure at room temperature. Thus, it can be injected directly into process streams operating near atmospheric pressure. The flow rate of ClCN is controlled using a regulator, and care must be exercised to ensure that the ClCN container maintains sufficient temperature to keep its total pressure above that of the process stream. The ClCN cylinder is commonly placed in a controlled heater bath to offset the cooling caused by the evaporation of ClCN during the test.

4.2.1.7 Solid-State Vapor Generator

The principal method of air purification for low-volatility vapors relies on physical adsorption of vapors onto high-surface-area adsorbent materials such as activated carbon. The solid-state vapor generator (SSVG) method reverses this procedure by initially contaminating a high-surface-area material with a small amount of the target chemical. Vapor generation is effected after purging the adsorbent particles containing the adsorbed vapor using dry air, nitrogen, or other carrier gas to reach a vapor loading equilibrium. A schematic depiction of this generation system is shown in Figure 4.8. This approach is highly effective in suppressing the equilibrium vapor pressure of the adsorbed material owing to the strong interactions between the vapor and the adsorbent. Studies have shown that the equilibrium vapor pressure of GB is suppressed by nearly 10 orders of magnitude using a mass ratio of 1% GB to coconut-shell carbon (CSC). The degree of suppression depends strongly on the mixing (vapor loading) ratio, and approaches the vapor-liquid equilibrium limit as the loading approaches the saturation limit of the adsorbent (0.4 g GB/g CSC). These observations immediately lead to the ability to precisely control the generation rate and concentration over an extremely wide range between direct generation of sub-parts-per-trillion concentration levels at very low flow rates. A generation rate of 100 femtograms (10^{-15} g) per minute for GB was observed. The generation rate measurements are limited on the low end by the detection technology used, not by generator conditions (loading ratio and temperature, primarily), which could have been further reduced.

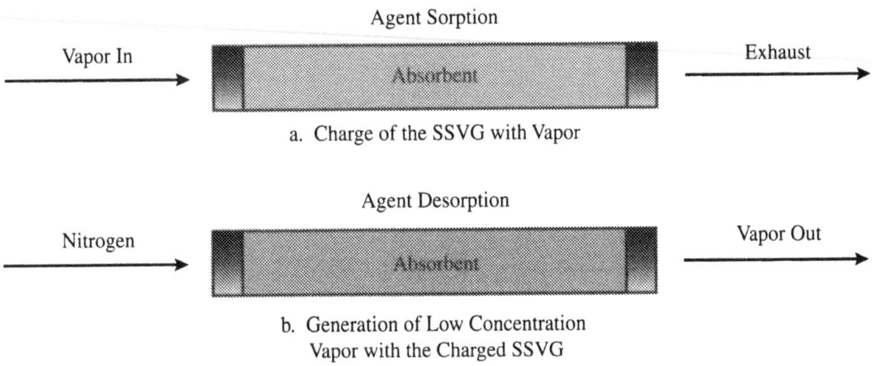

Figure 4.8 Solid-state vapor generator.

4.2.1.8 Compressed Gas or Gas Mixture

Using pressurized gases or gas mixtures is another way to generate vapors for TICs and other chemicals that are gaseous at room temperature. Pressurized gas can be purchased as pure gas or gas mixture at different concentrations. This gas can be blended directly with the dilution air in a dilution system to achieve the desired concentration. A special regulator may have to be used with each gas or gas mixture. When compressed cylinder gas is used, adjusting the flow rates from the pressurized vapor source and the dilution air stream governs the concentration. In general, because of the high volatility of this type of vapor, the concentration can be directly calculated. Adsorption losses can be considered minimal or negligible given their characteristic high volatility.

4.2.1.9 Chemical Reaction Method

Generation of chemical vapor through chemical reaction is an alternative method used to generate chemical vapors that are difficult to obtain commercially or are unstable under normal storage, or for other special purposes. For example, generation of a GB vapor can be achieved using the binary chemical reaction method. Vapor can also be generated via the decomposition of a chemical as well. The decomposing chemical serves as the source chemical. With this method, the emission rate is relatively easy to control using similar methods as discussed above, such as permeation and diffusion. For example, formaldehyde vapor could be generated using its water solution. But, water and methanol in the solution may interfere with the detection devices under testing. Thus, an alternate source of formaldehyde that would not interfere with the testing process must be used. The decomposition of paraformaldehyde releases formaldehyde without other byproducts. Therefore, formaldehyde vapor can be generated using paraformaldehyde contained in the permeation or diffusion device. The advantage of this formaldehyde source over the solution source is that it eliminates possible effects associated with water and methanol in the solution. Similarly, ammonia vapor could be generated using various ammonia-containing compounds, such as ammonia carbonate. The decomposition of ammonia carbonate releases ammonia as the source for generation.

The reaction of various chemicals is another way to generate chemical vapors. For example, SO_2 is released when HCl reacts with $Na_2S_2O_5$:

$$HCl + Na_2S_2O_5 = SO_2 + NaHSO_3 + NaCl \tag{4.8}$$

Difficulties associated with the reaction method are controlling the reaction that governs the release rates of the vapor, and the potential interference from multiple chemicals in the reaction during testing. Each and every reaction may require a special apparatus and settings.

4.2.2 Dilution and Mixing System

Having presented various techniques for providing a vapor source for generation, we will discuss ways of diluting vapor into the desired concentration and conditions for effective detector evaluation. Under normal circumstances, vapor concentration from the source can be accomplished with a dilution system to satisfy experimental needs. A single-stage dilution system is usually sufficient for most compounds to provide detectable concentrations for current detection devices. When very low concentration is desired, further dilution is required. A multiple-stage dilution system is then used.

Moisture in the air has been found to pose frequent undesired effects in many detection devices. Therefore, evaluating the moisture effect is one of the more important steps in evaluation. One way of mitigating the moisture effect is to include the humidity level in the generated vapor sample that is similar to the ambient condition. By so doing, the effect of moisture can be evaluated or negated. Therefore, the method for humidifying the vapor stream is presented along with discussion of dilution methods.

4.2.2.1 One-Stage Dilution System

A simple one-stage dilution system is shown in Figure 4.9. The vapor from the source, controlled with a regulated flow meter, meets the dilution airflow that is also controlled with a regulated flow meter into the mixing manifold. Diluted sample exhausts from the mixer can be used to test the detector directly or as a source for the next stage of dilution. The excess flow then passes through a charcoal canister to remove the toxic component before exhausted. The sample vapor concentration can be calculated as

$$C(\text{ppm}) = \frac{C_s(\text{ppm}) \times F_s(1/\text{min})}{F_s(1/\text{min}) + F(1/\text{min})} \tag{4.8}$$

where F_s is the flow rate of the source vapor, C_s is the source vapor concentration, and F is the flow rate of the dilution air. This single-stage dilution serves as the basic generation system from which later generator systems followed with added features. This dilution system is a simplified version leading to the two-stage dilution system described in the next section. The function of each component in this dilution system is similar to that in the two-stage dilution system, and is explained in the next section.

4.2.2.2 Two-Stage Dilution System

Figure 4.10 is an example of a two-stage dilution system. An air generator (6) provides the clean and dried air used for the carrier stream. This pressurized air is split into five streams, each controlled by flow meters equipped with appropriate needle valves (7, 11, 12, 19, and 20). Flow meters (11), (12), (19), and (20) control the two parallel humidity dilution streams. These are identical for ease of operation. The flow meter (7) is used to provide a dry airstream for the first-stage dilution.

For the two-stage dilution apparatus, the agent carrier gas (1) for the first dilution is supplied from a nitrogen tank and is controlled by a flow meter equipped with a

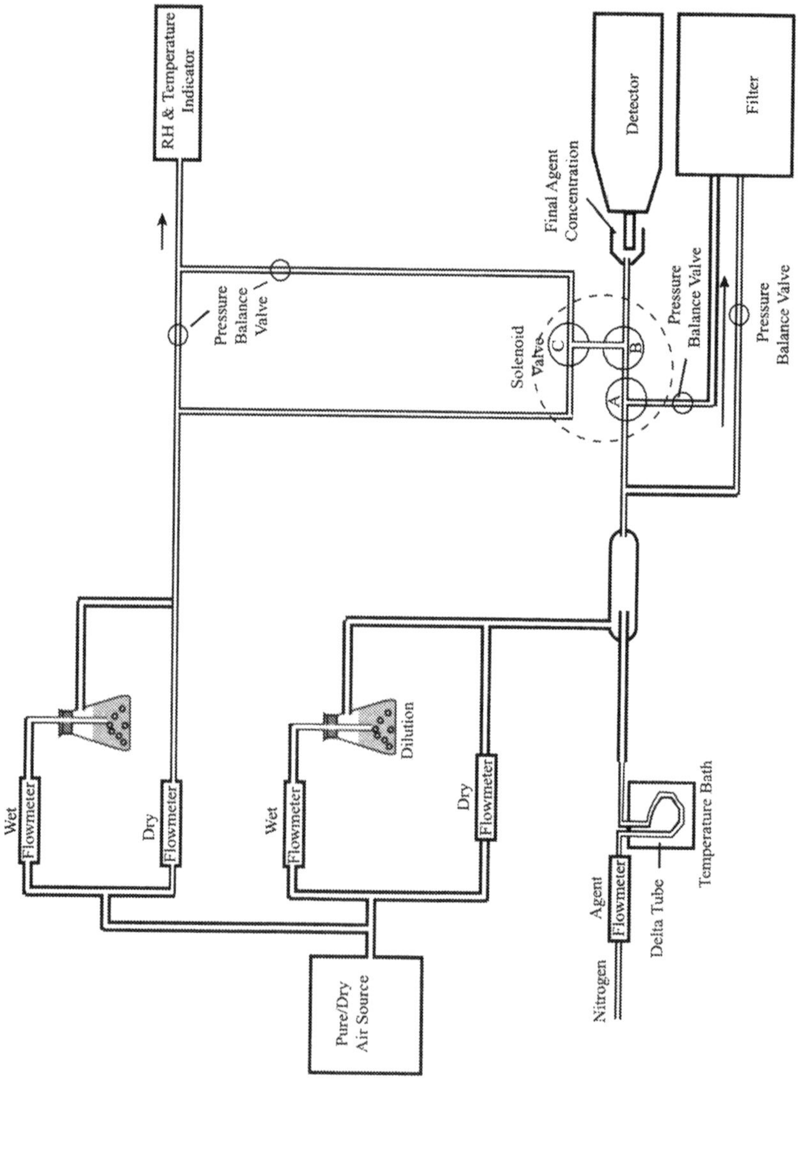

Figure 4.9 One-stage dilution system.

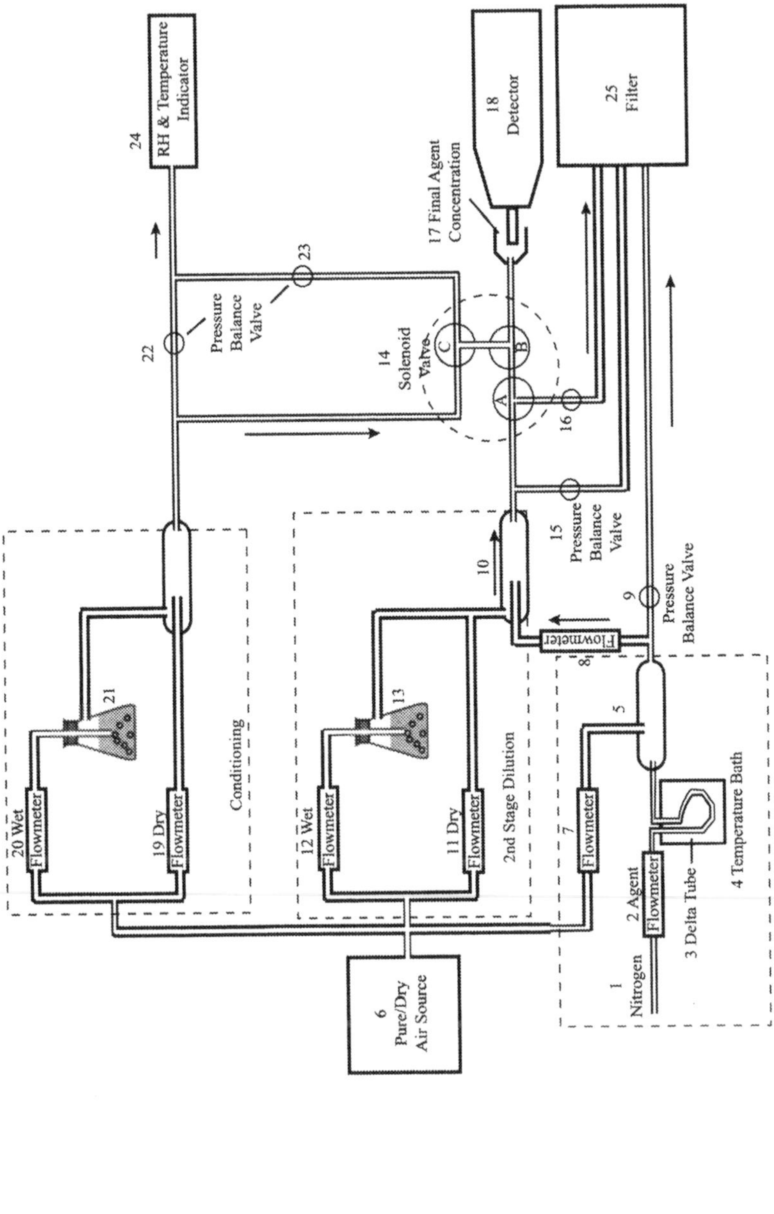

Figure 4.10 Two-stage dilution system.

high-accuracy needle valve (2) into the agent reservoir or the delta tube (3). The delta tube is immersed in a temperature-controlled bath (4) to maintain a constant temperature. The nitrogen carries the chemical vapor as it sweeps over the chemical and enters into a glass manifold (5) where it is met with a stream of dry air from the pure air generator through the flow meter (7) to provide the first-stage dilution.

A portion of the resulting first-stage agent vapor flow is forced through a flow control meter (8) and valve (9) into the manifold (10). It is met and further diluted with the humidity-tempered air from the combined streams from flow meters (11) and (12), as desired. The resulting relative humidity (RH) is controlled by varying the ratio of the dry air (0% RH) from flow meter (11) and the wet (100% RH) air stream from flow meter (12) bubbling through the humidifier water reservoir (13). The RH is estimated and is compared with the RH and temperature indicator (24) installed in the system's conditioning airstream. The humidified chemical vapor airstream is then split into two streams. One stream is directed to the solenoid valve assembly (14) consisting of three 3-way solenoid valves. The second stream is directed through a regulating valve (15) into a charcoal canister (25) to remove excess toxic chemical vapor.

A sample port (17) is connected directly to the solenoid assembly to minimize the delay when switching from conditioning air to agent air during a challenge. The glass cup has slots to release the excess flow that is delivered to the detector inlet area. The detector samples the agent vapor without having to attach directly to the sample port. Usually, the delivered flow is approximately twice the detector demand to ensure that the detector is not starved. This approach precludes detector response from pressure biasing.

This two-stage dilution system provides easily controlled, low-concentration level generation when limited airflow is required. It should be noted that the detector inlet is not hard coupled to the sample port and that the generated vapor is flowing over the inlet for the detector to sample; the flow must be in excess of detector demand. Total air delivered to the sample port must be minimized to ensure that toxic vapor released into the surrounding environment is minimal. The flow to the detector inlet, however, must be maintained in excess of detector demand. The reason for this type of sampling is to avoid generator flow from influencing the detection process through an induced pressure differential that could occur in directly coupled sampling.

Staged dilution extends the operating range to extremely low concentration levels while keeping total flow volume low. Presently, detection instrumentation limitations make it unnecessary to require more than a two-stage dilution for even the most volatile materials of interest. For example, GB is one of the most volatile liquid chemical agents tested routinely. Single-stage dilution can produce a final concentration near 0.1 mg/m^3 when the delta tube is held at 0°C at a total dilution flow rate of 3 l/min. Adding a second stage extends concentrations to below 0.0001 mg/m^3 while maintaining a final flow rate of 3 l/min.

However, if the concentration of the vapor source is very high and the target concentration is very low — for example, when a parts-per-billion level of Cl_2 vapor is generated with 99.99% Cl_2 gas — a multiple-stage dilution system may have to be used. A portion of the sample from the mixer of the first-stage dilution system is used as the source of the next stage of dilution and mixes with the second-stage dilution air similar to that demonstrated in the above example. A portion of this vapor is then diluted with anther dilution flow. This procedure greatly reduces vapor concentration and achieves the desired low concentration.

The following example shows the advantages of multiple-stage dilution over single-stage dilution. Let's assume that the starting concentration from a gas source cylinder is at a concentration of 1% and the estimated limit of detection is at the 1-ppm level. The generator system, therefore, must be able to generate vapor at that concentration level. Let's further assume for the sake of simplicity that the minimum reliable flow rate reading of the flow meter is 1 ml/min. To reach the 1-ppb vapor concentration using a single-stage dilution system would require a total flow rate (F_{total}) of

$$F_{total} = \frac{C_s \times F_s}{C_r} = \frac{1\% \times 1(\text{ml/min})}{1(\text{ppb})} = \frac{10,000(\text{ppm}) \times 1(\text{ml/min})}{0.001(\text{ppm})} \tag{4.9}$$

$$= 10,000,000(\text{ml/min}) = 10(\text{m}^3/\text{min})$$

Therefore, the required dilution flow is 1 ml/min less than 10 m^3/min. Such a large amount of flow, greatly in excess of detector needs, creates undesirable side effects. Because CWAs or TICs are highly toxic, their release into the atmosphere should, if possible, be kept to a minimum to avoid excessive contamination. Therefore, the excess exhaust from the generation process must be contained via trapping or filtering through an activated charcoal filter to capture toxic material from the stream before it is exhausted.

A flow of contaminated air at 10 m^3/min has created safety concerns when such a large volume of toxic vapor is being continuously created. Higher volume usually leads to higher system pressure, more toxic substance being consumed, large-area contamination potential, and a larger apparatus requirement. All of these factors create additional hazards for operators. This flow cannot be easily controlled or disposed of through a small-size filter in the laboratory setting. Use of large filtering system drastically increases operational logistics and costs.

Furthermore, the use of large volumes usually increases the system's internal pressure. Consequently, the flow through system plumbing could create undesired artifacts when sampled by the detection device during testing. The internal pressure effect could become drastic when the detector is directly connected to the sample port. Such effects may drastically affect the detection algorithm of many detectors. A preferable method is to use a smaller amount of excess vapor around the detector inlet through a sample cup so that the detector will draw in the amount it needs without being influenced by the relative flow velocity from the generator system.

When a two-stage dilution system is used, the concentration is first reduced to, for example, the 3-ppm level by using 1-ml/min source vapor and 2999-ml/min dilution air:

$$C_1 = \frac{10,000(\text{ppm}) \times 1(\text{ml/min})}{3000(\text{ml/min})} = 3.3(\text{ppm}) \tag{4.10}$$

We can, then, take 1 ml/min of this diluted vapor and dilute it with again with 3000 ml/min of dilution air to reach the desired 1 ppb second-stage concentration in a total flow of 3 l.

$$C_2 = \frac{3.3(\text{ppm}) \times 1(\text{ml/min})}{3000(\text{ml/min})} = 1.1(\text{ppb}) \tag{4.11}$$

This total 3-l/min flow can be controlled much more effectively using a small filter. Thus, the two-stage generator provides an effective means to produce a desired concentration of these toxic materials for laboratory testing purposes.

4.2.2.3 Other Generation Techniques

Besides the discussed methods of vapor dissemination for subsequent dilution to achieve the desired agent concentration, there are several additional techniques that are worth mentioning. Vapor generation from low-volatility compounds may not provide a sufficiently high concentration using the delta tube, diffusion, permeation tube, or solid adsorbent techniques. The use of a syringe pump and subsequent evaporation at a small flow volume does not produce easily controllable concentrations in the laboratory. To overcome such problems and provide an adequate, stable, generated concentration, an aerosol generator would be necessary.

An aerosol generator, unfortunately, will produce aerosol or particles together with a vapor. In the case of low-volatility liquids or solid compounds, such a practice of producing a test concentration is considered acceptable. More than likely, the dissemination of low-volatility compounds would be in the form of aerosols to produce the desired concentration. In the case of low-volatility liquid compounds, the use of a nebulizer to produce a spray of fine mist for subsequent dilution can produce the desired concentration. Figure 4.11 presents a flow schematic for a typical aerosol generation system. A push-pull system is required to aid aerosol movement along the tube. Based on the nebulizer, dilution flow rates, and viscosity relationships, the lighter aerosols will be carried into the dilution stream while the larger sprays will hit the wall of the nebulizer envelope and flow back into the liquid.

In addition to the use of a nebulizer, another specially designed apparatus has been used for generating specific concentrations from solid compounds is a vapor condensation aerosol generator (Figure 4.12). The solid is melted in a reservoir immersed in a heated silicone oil bath controlled at above the melting point of the substance. Similar to vapor generation, the nitrogen stream passes through the molten substance to carry the vapor as well as the aerosol produced by the bubbling effect to meet with the dilution airflow. The dilution airflow can be conditioned with proper humidity, as in the case of delta tube vapor generation. The cooler dilution air will condense the hot vapor and form aerosols of fine solids. Depending on the flow of the carrier and the dilution, a desired concentration can be achieved. The heavier particles that were not carried to the end of the vertical tube because of the flow relationship will fall back into the agent reservoir where they are recycled.

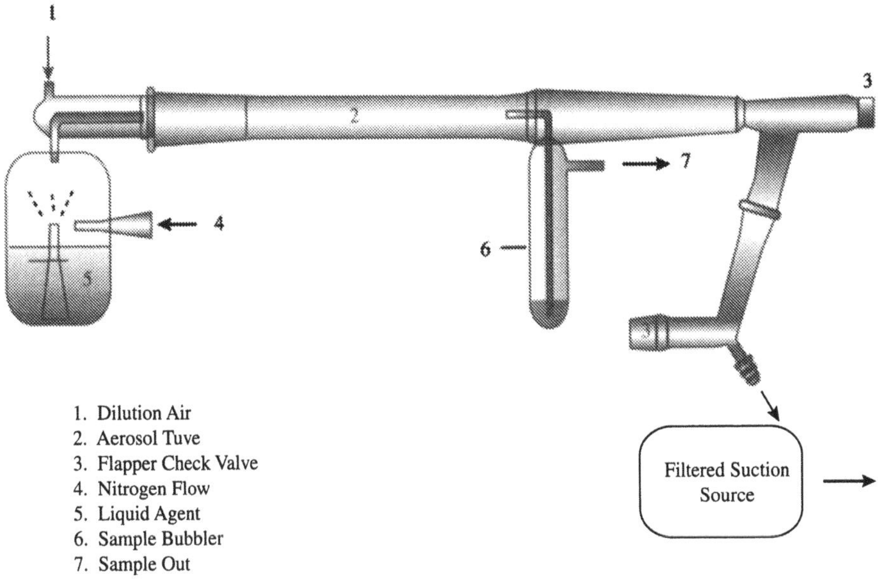

Figure 4.11 Liquid aerosol generator.

To maintain the flow of aerosolized particles from dissemination toward the sample port and eventually the exhaust, it is necessary to provide a suction source at the end of the tube to equate the amount of air used for the generation process. This balancing of input and exhaust facilitates aerosol movement through the tube, which prevents aerosol fallout due to generator's internal pressure resistance that causes droplet coagulation.

4.2.3 Comparison of Generation Techniques

Spargers have been widely used for generating a high-flow airstream. This method has general utility, but is subject to the unintended release of liquid in the event of pressure transients downstream of the generator. This possibility presents a serious problem for generation of high-toxicity vapors, and extraordinary safety measures (e.g., pressure relief valve and liquid trap) must be used to preclude liquid release when used for CWA vapor generation.

Another difficulty experienced when using the sparging method is the possibility of unintended aerosol production and transport with the vapor. Expedient measures can be implemented to remove aerosols from the product stream, including aerosol filtration and maintaining the sparger at subambient temperature; however, this phenomenon reduces the operator's ability to establish predictable concentration with the generator's operating conditions.

The use of a delta tube saturator and the two-stage dilution system described above has the capability of generating highly purified agent vapor in a small controlled volume of conditioned air within the limits imposed by their physical properties. The small agent quantity permits rapid purging of the higher-volatility impurity materials that may interfere with generation of the target vapor. This is the only

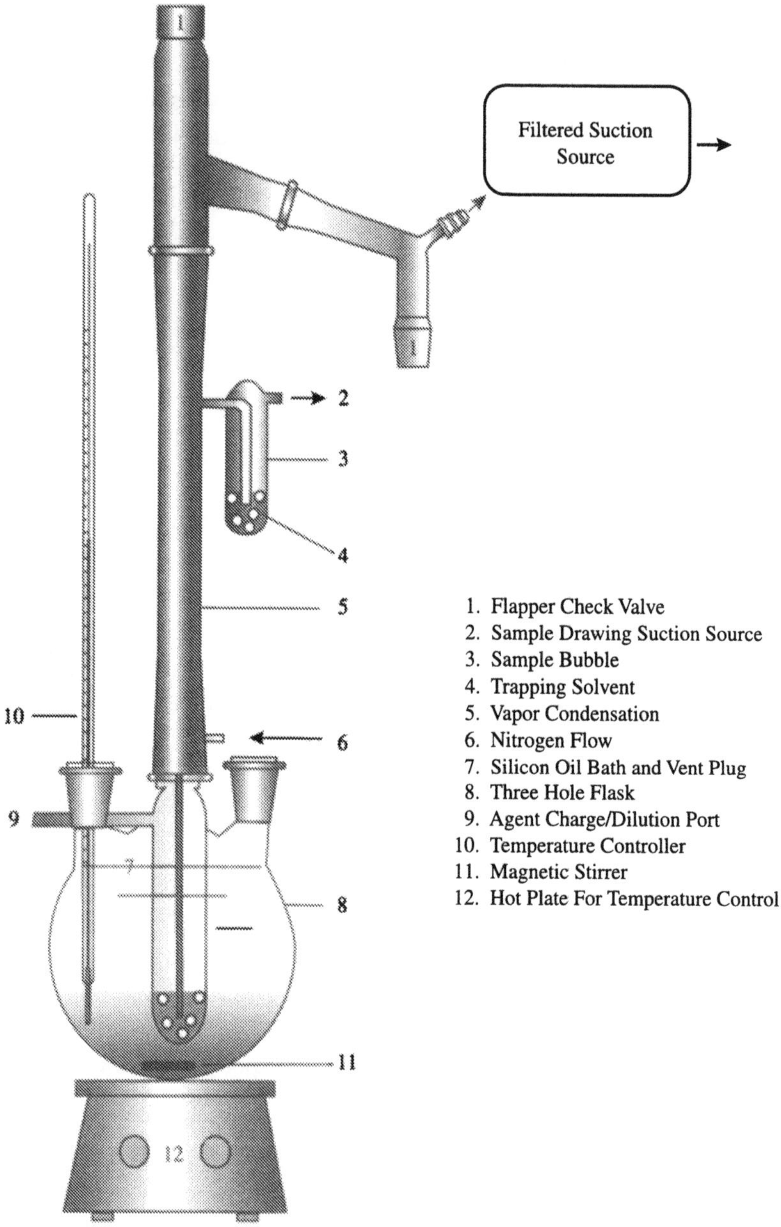

Figure 4.12 Solid aerosol generator.

method that can produce highly purified target vapor where highly volatile impurities cause interference.

The delta tube behaves similarly to the diffusion generator except that it has a more positively controlled carrier flow that can overcome the generator's internal system pressure. A diffusion system, which the vapor from the substance passively diffuses depending on flow dynamics of the system, can be quite unpredictable due to changes in internal pressure. The flow rate of the carrier gas in the delta tube generation system can be precisely controlled using a needle valve. More importantly, the use of the delta tube prevents dangerous backsplashing of the agent that can occur when an impinging sparger is used improperly. Backsplashing is an area of major safety concern when dealing with lethal substances.

The use of a two-stage dilution system expands the range of achievable concentrations while keeping the final dilution flow rate to a minimum as required by the experiment and safety considerations, without having to resort to high-volume dilution to reach low-concentration generation. The two-stage system also facilitates generation and control of RH and vapor concentrations. Only the last-stage RH needs to be controlled. Raising or lowering the final concentration can be achieved by varying the agent flow rate of the first stage, while keeping the same second stage flows, therefore, retaining the same RH condition in the final stage.

Saturator systems can produce the desired concentration very rapidly with minimal manipulation. Because pure substance is used, the resulting vapor is of the highest purity without solvent influence. In cases where higher generation rates are needed, a high-capacity multiple-pass saturator (Figure 4.4) or parallel addition of delta tubes (Figure 4.13) together with higher bath temperature can be implemented.

Syringe pump methods work well and are generally applicable for a wide variety of materials. However, this method has disadvantages and limitations that should be recognized by the potential user. Containment difficulties arise for low-boiling-point liquids. Also, the ability to generate vapors at very low rates is limited currently to greater than 0.1 mg/min or 0.1 ml/min by available equipment. Additionally, since this method involves liquid injection, care must be taken to ensure that complete evaporation and mixing are achieved prior to sampling the product stream. Cooling due to liquid evaporation can be an exacerbating factor for the latter, particularly at high injection rates owing to the low heat capacity of air. For example, injection of liquid water into a dry airstream at 25°C to achieve a 10% RH product will result in an airstream at approximately 19°C and 15% RH unless the stream is reheated to restore the original temperature. This effect will be less significant for low-humidity, low-flow conditions, since the heat of evaporation will be low. The effect will become more pronounced at high-flow, high-humidity conditions.

Permeation tube generators are predictable, safe, and convenient to operate. The two principal disadvantages are low generation rates and long equilibration time as compared to delta tube methods. Applications are limited to a smaller range of compounds. VX, for example, has not been successfully generated using a permeation tube owing to effects caused by impurity vapors that interfere with VX permeation. Impurities within the VX, although in very small percentages as a liquid, have much higher vapor pressure values than VX. This vapor pressure phenomenon greatly altered the ability to effect proper VX permeation. The higher vapor pressure impurities that permeate at faster rates thus overwhelmed VX concentration. Only the delta tube method is capable of producing high-purity VX vapor.

The SSVG method is ideally suited for direct generation of low flux streams (i.e., low concentration and low flow rate) without the need for further dilution. Another potential advantage of this method, which has not been explored yet, is the ability to reduce head pressure over a high-volatility vapor or gas that might normally

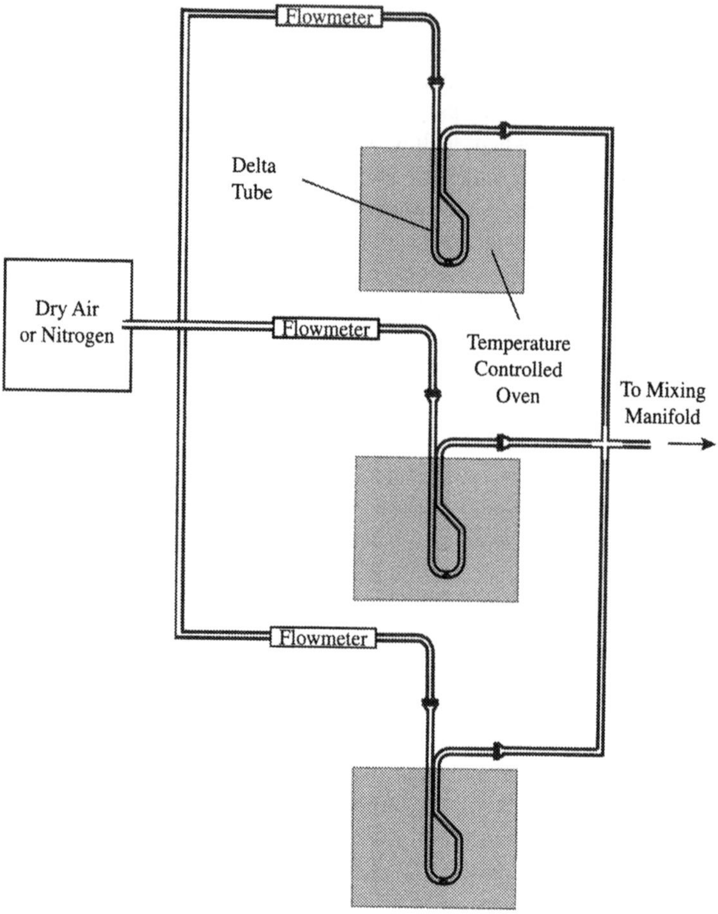

Figure 4.13 Setup of parallel delta tubes.

be dispensed from a high-pressure container. This safety feature might be particularly useful for field dissemination of CWAs such as cyanogen chloride and arsine for detector calibration, but its greatest utility should be realized in industrial applications such as gas (hydrogen) storage. The wide range of available generation rates and control parameters using this method make it ideal for a wide range of applications. In addition, storage of highly toxic vapors on a solid adsorbent can significantly reduce the safety risks associated with handling these materials.

Potential problems associated with the SSVG method include decomposition of the adsorbate as a result of intimate contact with adsorbent reactive sites. This effect has been observed for adsorbed HD and VX recently, and seems to be exacerbated by the presence of water vapor. Efforts to minimize this effect include use of low-ash adsorbents, operating at as low a temperature as possible, and avoiding the presence of water. The latter includes sample preparation methods to reduce water contamination and use of ultra-dry carrier streams since water vapor in the carrier even in the parts-per-million range will eventually be concentrated on the surface of the adsorbent and potentially lead to hydrolysis.

The high-pressure vapor injection method has the distinct advantage that a known dose of chemical is delivered to the test rig. Also, the integrated concentration time (Ct) challenge can be determined if the flow rate is known. Often, however, test planners desire step function challenges, that is, rapid establishment and removal of the feed concentration over the test duration. In the present case, the feed concentration rises rapidly but tails off slowly due to mixing effects inside the mixing chamber. On the other hand, this injection method more closely approximates what might be expected under field dissemination. In fact, mixing effects in the field will most likely be more pronounced than observed in the test configuration.

Under suitable conditions where rapid change of generated concentration is not required, permeation tubes can be used as the source of low concentrations of gases and vapors. Once the system becomes equilibrated, the same concentration can hold for a long period of time. The disadvantage is the longer lag time for achieving different concentrations. Thus, this method of generation would be best suited for routine testing using the same concentration.

Syringe pump generation has a low-concentration generation limit, injecting as little as 0.1 mg/min of GB into an airstream at a flow rate of 1000 to 1500 l/min to reach a final concentration near 0.1 mg/m^3. It is not suitable for low-concentration generation.

4.2.4 Humidification of Generated Vapor

Generating humidified vapor streams using air as the carrier to simulate an operational environment for detector testing is a must. Moisture often causes changes in detector behavior. In this case, an added complication arises due to the interaction between water vapor and the target vapor. Ideally, Raoult's Law limits the sum of relative pressures (ratio of partial pressure to single-component vapor pressure at the measurement temperature) of all vapors in a mixture to 1.0. For example, an air stream at 50% RH can only hold up to 0.5 relative pressure of the target vapor. Deviations from Raoult's Law can be significant, especially for mixtures involving nonmiscible liquids.

As mentioned earlier, detection devices should be tested under diverse environmental conditions. Therefore, the basic generator system was modified to include humidification of the vapor stream. Since moisture will degrade the source substance, the carrier stream passing over the substance must be kept at zero moisture content to prevent sample degradation. Thus, only dilution air streams can be humidified. Fortunately, the amount of carrier flow through the source is much less than the amount used for subsequent dilution, except for very low volatility compounds. That permits generation of a vapor stream with controlled RH.

The one- and two-stage dilution apparatuses discussed earlier include the humidifier devices shown in Figures 4.9 and 4.10. The following humidifying process discussion is based on the more complete system shown in Figure 4.10. The conditioning side of the generator described serves as an example of a humidifying method. The humidifying stream uses the flow from flow meter (19) that is set equal to the combined flows through flow meters (8) and (11). This dry airstream is combined with the flow from flow meter (20) that is set equal to the flow from flow meter (12), through the humidifier (21), to form the conditioning air stream. This stream is also split into two streams similar to that described in the chemical vapor generation section. One stream passes through the regulating valve (22) and is released. The other stream is directed to solenoid (B), and through solenoid (C) into the sampling cup (17), while the solenoids are not energized. Challenge with the chemical vapor is effected when all three of the solenoid valves (14) are energized. The conditioning air stream exhausts through solenoid (C) and a regulation valve (23) to the RH and temperature indicator (24) for an indication of the RH and temperature conditions. This arrangement permits a wider range or regulation of flows to meet test demands. Because flows for controlling humidity of the two sections are matched with identical flow rates, readings obtained from the humidity indicator can be considered as equal to the RH condition of the chemical vapor-generating section.

To ensure that the air is saturated when it passes through the water reservoir kept at the same temperature as the environment, the flow is maintained at the minimum level to provide adequate volume (e.g., 3 l/min is adequate for most detectors). Saturation efficiency needs to be considered when higher flows are used, however. By keeping the humidifier water at the same temperature as the environment, the relationship will hold. The air passing through the humidifier can only become saturated at the same temperature and thus, when it is mixed with the dry air, the percentage of RH is directly proportional to the ratio of wet and dry flows. This technique minimizes the need to manipulate flow adjustments to control RH if and when a heated water source is employed. It also eliminates condensation problems associated with heated water or steam injection techniques.

While the described generation system uses dual humidified streams, the humidity of the toxic vapor stream is not measured. The RH is considered the same, for practical purposes, as that of the conditioning side by use of similar flow settings. The RH on the conditioning side is measured when detector testing is underway. Quite often, detection sensors have been found to be grossly affected by humidity differences between the conditioning and the challenge airstreams. RH mismatch could cause large changes in baseline signals. Therefore, it is necessary to provide identical RH to both the conditioning and subject vapor streams. To ensure that there is no RH deviation between the two streams, the generation apparatus is modified to use a single humidifier rather than two separate humidifiers (Figure 4.14).

4.3 GENERATION OF CWA OR TIC VAPOR WITH INTERFERENT VAPOR

The ban against outdoor testing using toxic chemicals prompted the need to generate toxic vapors together with a potential interfering vapor under laboratory engineering controls for detector testing. In general, the potential interference vapor is introduced to the vapor generation system immediately before the vapor stream enters the sample cup. The interference vapor source can be disseminated through

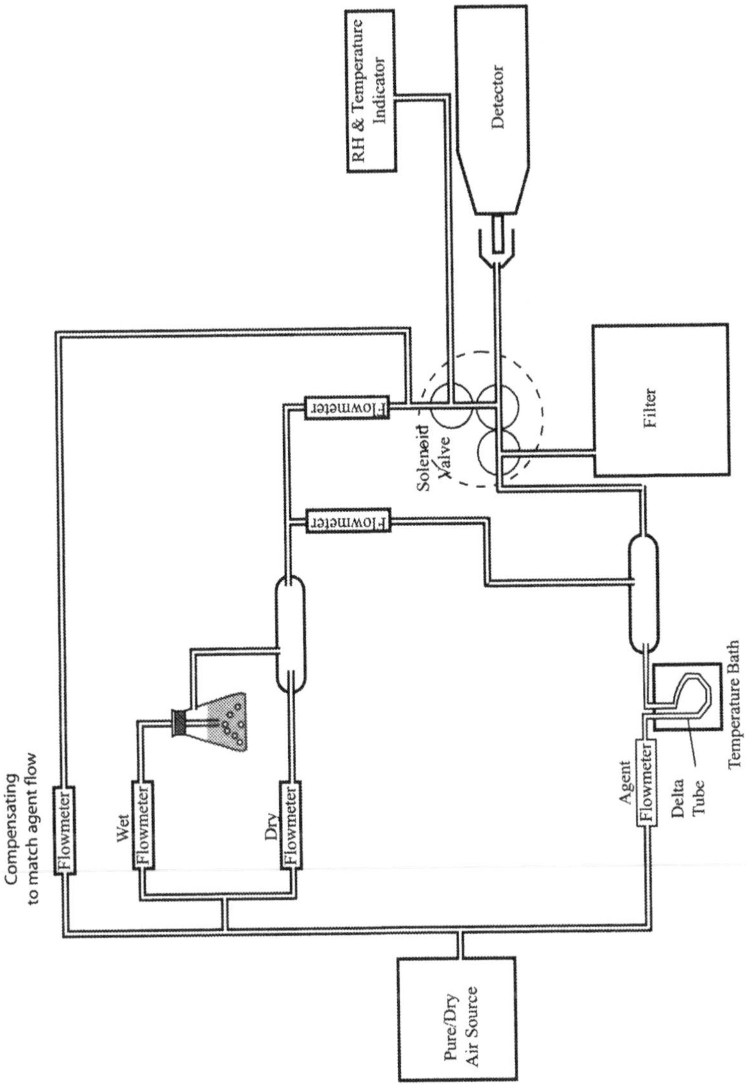

Figure 4.14 Single humidifier for condition and agent generation.

several methods, including delta tube, impinging bubbler, compressed gas cylinder, engine exhaust, and burning apparatus.

The flow passing through the delta tube or impinging bubbler containing the liquid interference sample is controlled to carry the saturated headspace concentration of the substance. Figure 4.15 shows the diagram of a two-stage dilution system coupled with a gas dispersion bubbler that generates necessary interference. In general, passing the air through the liquid in a bubbler is an efficient way of saturating the air with a given substance. However, many substances foam readily by the bubbling action. A delta tube can then be used so that the air only sweeps above the liquid surface to carry the headspace vapor for dilution. Depending on the total flow of the basic vapor generation system, the flow through the interferent delta tube or impinging bubbler is adjusted to produce the desired volume dilution. For example, when the final flow of the generation system is set for 3-l/min, the amount of flow through the interferent will be 30 ml/min to achieve the 1% headspace concentration.

Similarly, the flow from a compressed cylinder can be properly converted to provide the desired parts-per-million concentration when diluted with the flow from the generation system (Figure 4.16). Gasoline engine and diesel engine exhaust can be piped into the laboratory and the flow can be controlled by use of a stopcock to give the proper volume dilution.

To generate smoke from burning materials requires much more manipulation. Figure 4.17 shows a laboratory smoke generator. It is necessary to provide airflow to support the combustion necessary to properly burn the sample (e.g., with diesel fuel, kerosene, jet propulsion fuel, etc.). A small amount of the sample is placed in the cup attached to the center stem. A cotton string, used as a wick, is lit. The fixture is inserted into the flask where a stream of air is provided to maintain combustion. Once the proper combustion is achieved by regulating the amount of air into the flask, the flame is extinguished. The stopcocks are then adjusted to force the desired volume of air toward the vapor generation system while exhausting the rest of the combustion air. Once this relationship has been reached, the wick is again lit and testing can then begin.

To test the effects of the interferent on the detection ability of a device, the following procedure can be adopted. A generation system equipped with parallel conditioning and vapor streams connected to three solenoid assemblies is used. The interferent stream mixes with the stream from the solenoid assembly prior to the sample cup where the detector draws in the vapor sample. The detector is first being conditioned by sampling the conditioning air. After a period of conditioning, the solenoids are energized and the detector then draws its sample from the vapor stream to ensure that it detects the target substance of interest. These took place before the interference is connected or introduced.

Once the detector has shown proper response to the subject vapor, it is allowed to clear under the conditioning air. After clearing, the interference vapor is introduced and mixed with the conditioning air to observe the effect on the detector. If the detector responds with an alarm to this interference vapor, it could be misconstrued as the presence of the target toxic chemical.

If exposure to the interference vapor does not cause a false positive response, the conditioning air is replaced with the subject vapor air. The detector is now

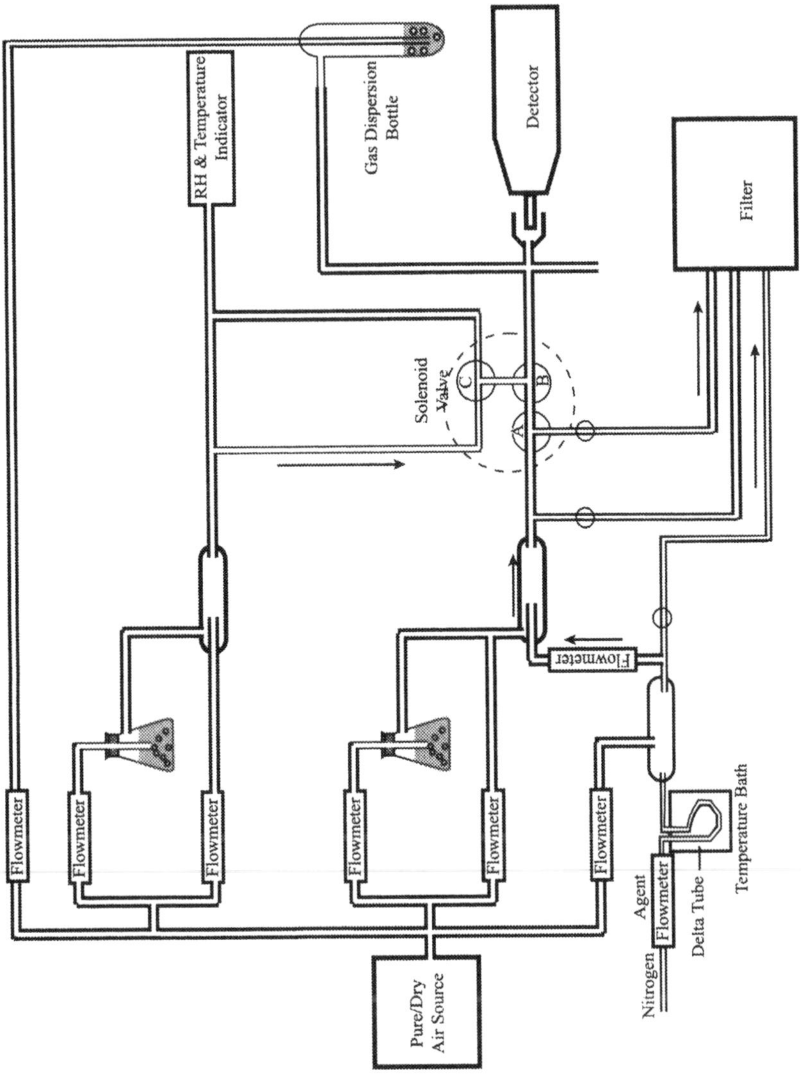

Figure 4.15 Two-stage dilution system with gas dispersion bottle as interference source.

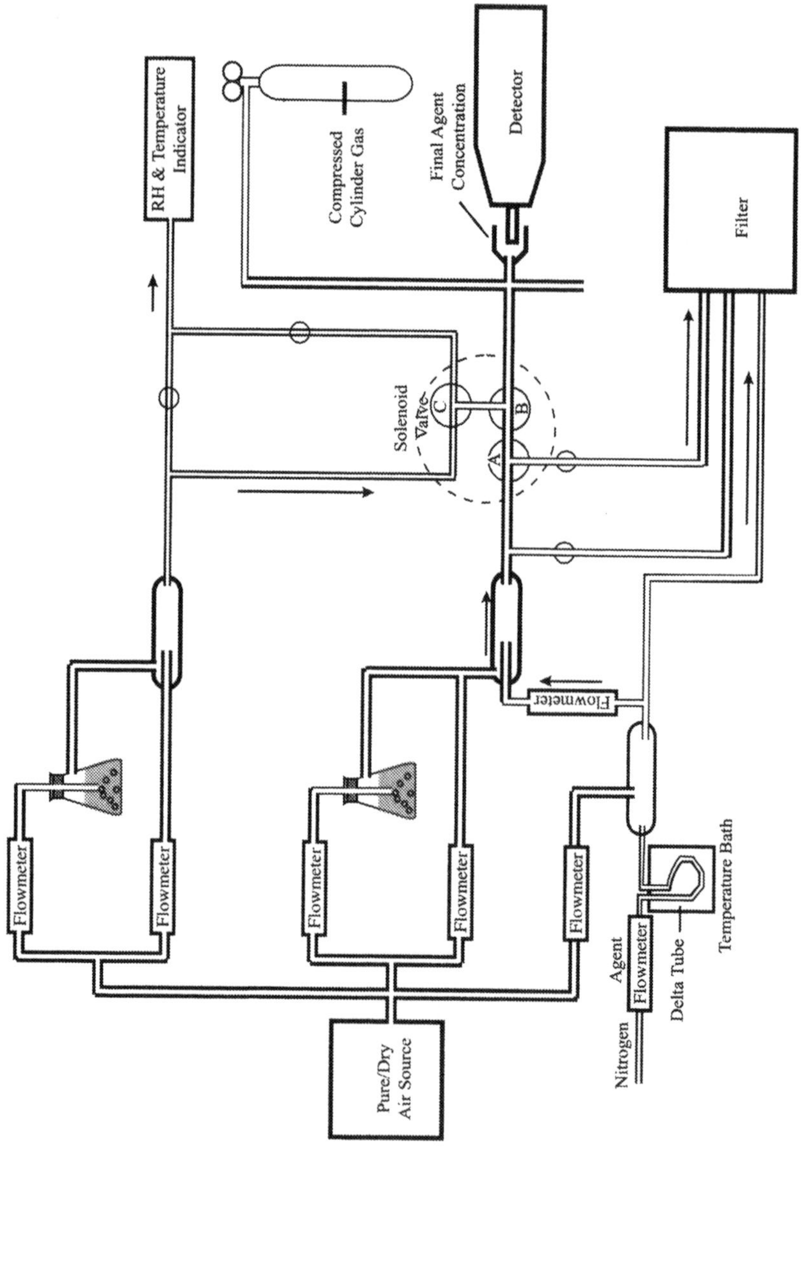

Figure 4.16 Two-stage dilution system with compressed gas as interference source.

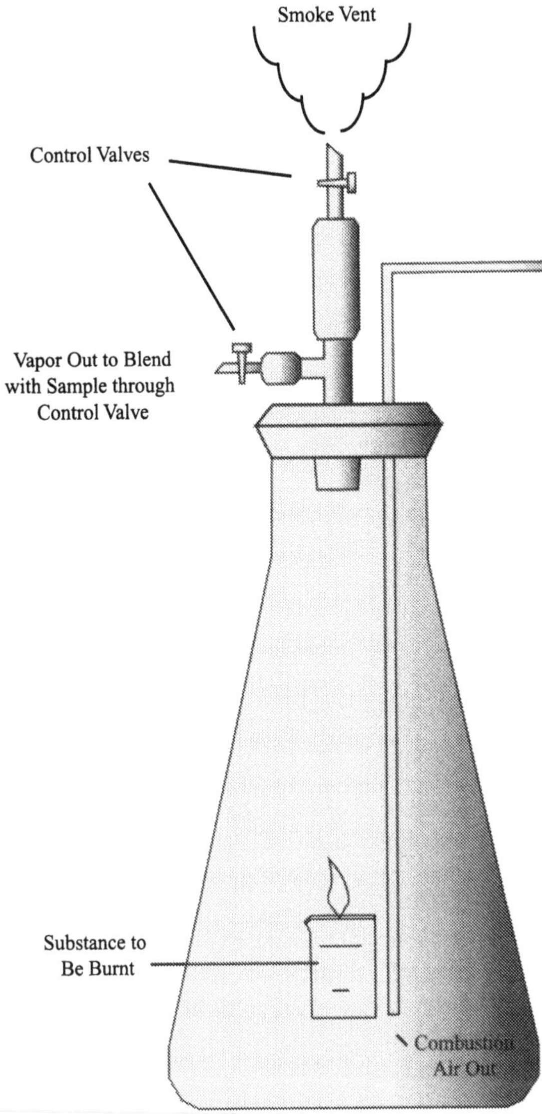

Figure 4.17 Smoke generator.

sampling the subject vapor together with the interference vapor. If it can properly respond to the subject vapor, the detector is considered not affected by the presence of the interference vapor. If, however, the detector fails to respond with an alarm, the effect is considered false negative; although the interference vapor does not cause the detector to generate false positive alarms, it interferes with the detection algorithm to prevent proper detection of the target vapor. This effect is considered more harmful than a false positive response. While false positives are a nuisance, false negatives could be fatal.

Laboratory testing is limited due to a variety of restraints. For instance, laboratory equipment often creates undesired artifacts that produce misleading results.

Mismatches in RH and pressure influence during sampling are the most common artifacts. Care must be exercised to identify and minimize or eliminate such artifacts during any evaluation. Rigorous evaluation of toxic chemical detection devices is crucial to ensure that they are reliable, that is, that they meet their design objectives.

4.4 CWA SIMULANTS

Detectors and protection equipment should be tested with actual CWAs or TICs during the development stage to ensure performance. While it is possible to obtain various TICs from commercial vendors, tests with actual CWA agents are generally limited to a few surety-qualified laboratories and government agencies. Open-air tests with CWAs have been banned since the sheep killing incident in Utah linked to VX leakage in 1969. Hence, CWA simulants are used as alternatives. Studies have been conducted to evaluate the properties of protective clothing with simulants. (Reports can be accessed at http://hld.sbccom.army.mil/ip/reports.htm#detectors or www.cdc.gov/niosh/nppt1/simulppjul31.html.)

To be qualified as a CWA simulant, the chemical should mimic the behavior of the CWA it simulates. It must be relatively nontoxic if the simulant is intended for outdoor dissemination. The chosen substance must be environmentally benign. If the simulant is intended for use under laboratory control conditions, substances with some toxicity can be used. Simulants are chosen based on structural, vapor pressure function group similarities to respective CWAs. When the simulant is used to evaluate detector performance, choosing the simulant that closely matches the type of detection methodology used by the detector is of the utmost importance. For example, if the detection device is an ion mobility detector, the simulant chosen must have similar ionization affinity, function group, and structure size so that its mobility is similar to the CWA.

Tables 4.3 and 4.4 list several substances that have been used to simulate the nerve agent GB and blister agent HD, respectively, in detector testing. The listed chemicals in Table 4.3 are similar to GB in some aspects. For example, the structure of dimethyl methylphosphonate (DMMP) is very similar to that of GB. Both contain the CH_3-P=O group. The main difference between them is that the simulant does not have the more active fluoride (-F) function group, and thus, its toxicity is much lower than GB. Both compound molecules contain phosphorous. Therefore, it is possible to use DMMP as a simulant to evaluate the performance of a flame photometric detector, ion mobility detectors, or surface acoustic wave detector.

Table 4.4 lists various chemicals that have been used to simulate the behavior of mustard gas. For example, dibutyl sulfide (DBS) has a similar structure, and is used as a simulant of mustard gas for detector evaluations.

However, no matter how closely simulants mimic respective agents, they are not the real thing. Simulants should only be used in preliminary studies. For example, during development of a flame photometric detector, DMMP and DBS can be used to generate the necessary signals for the flame photometric detector. DMMP contains phosphorous and DBS contains sulfur atoms; therefore, they are suitable for simulation of the G and H agents, respectively. The use of these

Table 4.3 GB and Simulants

Chemical	Structure	CAS	Notes
GB	$CH_3P(O)(F)OCH(CH_3)_2$	107-44-8	
Dimethyl methylphosphonate (DMMP)	$CH_3P(O)(OCH_3)_2$	756-79-6	Vapor pressure (mmHg) = 1.2 mm Hg at 25°C
Diethyl methylphosphonate (DEMP)	$CH_3P(O)(OC_2H_5)_2$	683-08-9	
Diethyl ethylphosphonate (DEEP)	$C_2H_5P(O)(OC_2H_5)_2$	78-38-6	
Diisopropyl methylphosphonate (DIMP)	C_7-H_{17}-O_3-P	1445-75-6	Vapor pressure = 13 mmHg at 10°C. Diisopropyl methylphosphonate is a chemical byproduct resulting from the manufacture of sarin (GB).
Triethyl phosphate (TEP)	$(C_2H_5O)_3P(O)$	78-40-0	

Table 4.4 HD and Simulants

Chemical	Structure	CAS
Bis(2chloroethylethyl)sulfide (HD)	$Cl(CH_2)_2S(CH_2)_2Cl$	505-60-2
Dibutyl sulfide (DBS)	$CH_3(CH_2)_3S\ CH_3(CH_2)_3$	544-40-1
1,6-Dichlorohexane (DCH)	$Cl(CH_2)_6Cl$	2163-00-0
Di-n-butyl disulfide (DBSS)	$CH_3(CH_2)_3S_2(CH_2)_3CH_3$	629-45-8
Bis 4-chlorobutyl ether (BCBE)	$Cl(CH_2)_4O(CH_2)_4Cl$	6334-96-9
2-Chloroethyl phenyl sulfide (CEPS)	$C_6H_5SCH_2CH_2CL$	5535-49-9

chemicals for development purposes will make the process easier and more efficient because of the limited CWA availability for testing. Tests with actual CWAs must be performed periodically during developmental stages. Occasionally, the results observed in testing using simulants do not match those derived from using the actual agents.

A simulant that is acceptable for one purpose may not be useful under different conditions. Light emission from the excited S_2 through burning of dibutyl sulfide (DBS) makes it a suitable candidate to serve as a viable HD stimulant for flame photometric detectors. If a device is designed for detecting Cl clusters formed by decomposing HD molecules, the use DBS as an HD simulant will not work. There is no chloride contained in DBS. Another substance that contains chloride, such as CEES, would then be proper. Therefore, deciding on the correct simulant requires a thorough understanding of the unique characteristics of the target CWA and potential simulants.

It is very important to note that although the toxicity of the chemicals listed in Tables 4.3 and 4.4 is much lower than CWAs, some of them are still quite toxic and are considered controlled substances. For example, diisopropyl methylphosphonate (DIMP) is a byproduct of GB production. Before using any of the chemicals listed in the tables as a simulant in a given situation, substance toxicity and impact on the environment must be considered. The U.S. Army Edgewood Biological and Chemical Center maintains an extensive database containing structural, physical, chemical, toxicological properties, and applications information on chemicals that have been tested or proposed for use as CWA simulants.

CHAPTER 5

Detector Selection Factors

To combat terrorism, we must anticipate what terrorists might do. Given that the CWAs are under tight controls, terrorists may resort to compounds that may be less toxic than the CWAs but that are easily accessible in large quantities such as the TICs discussed in Chapter 2. Thus, a single instrument (or small number of instruments) that can detect TICs as well as the more toxic CWAs is desirable.

In later chapters we discuss principles of operation and advantages and disadvantages of each type of detection technique. There is no perfect detector that could possibly satisfy all requirements, at least given current state-of-the-art technologies. We hope that our review will assist readers to select appropriate detection devices for their needs from the vast number of commercial off-the-shelf offerings.

Currently, available detectors vary in cost, performance, and reliability. To choose the correct device for an operation can be quite challenging. While cost may be a concern, choosing the best and most reliable detector for operational needs is more important. Although many existing detectors have been tested under diverse test programs, both governmental and industrial, many others have not been tested. Many detector manufacturers make claims based on their own testing. Some of those claims have not been thoroughly verified by third-party laboratories. Therefore, care must be taken in digesting available data on device performance and testing.

Many factors should be considered when selecting a detector, and we will discuss several important ones. There is no universal set of factors that will fit every objective. Depending on the specific application and circumstances, a detector that is ideal for one type of operation may not be appropriate or the best choice for another. For example, formaldehyde or other volatile organic compound (VOC) detectors may not be suitable for detection of nerve agents. A detector that is capable of detecting nerve agents may not be very useful for detecting certain TICs. To choose a suitable detector, knowing the target chemicals is important, as this information will help to determine what types of techniques are best for the application. In subsequent chapters, we explain several prominent detection techniques and the chemicals that they can detect.

Two types of factors should be considered in selecting a detector: detection capability (selectivity, sensitivity, response time, etc.) and performance (warm-up time, calibration requirements, portability, and power requirements) factors. These factors are discussed below.

5.1 SELECTIVITY

"Selectivity" is the term used to describe the ability of a detector to respond only to the targeted chemicals in the sample. When a detector is set to detect G-agents, it should only respond with G-agent detected in the sample, regardless of how many other chemicals are present. If, for example, pesticides in the sample interfere with G-agent detection, selectivity would be regarded as less than perfect. Depending on the techniques used, some detectors have better selectivity than others. For instance, flame photometric detectors respond only to phosphorus and/or sulfurous compounds through the use of specific light spectra, while flame ionization detectors are better equipped for detecting organic compounds. Given that most CWAs of interest contain either phosphorus (nerve agents) or sulfur (mustard gas), flame photometric detectors would serve well as a specific CWA detection device.

Colorimetric types of detectors are more specific to certain compounds through various chemical reactions. Therefore, they are highly specific detection devices and have higher selectivity, as they respond to a certain number of specific chemicals that they were designed to detect with little or no interference from other chemicals. Consequently, the potential for false responses/alarms is limited. More selective detectors are limited in terms of the number of compounds that could be added to their detection capability. Colorimetric detection devices would be a good choice for monitoring known chemicals in the air at a site where the specific chemical of interest may be present, such as monitoring the phosgene exposure level while personnel is working with phosgene in the laboratory. Because colorimetric detectors are highly selective, multiple colorimetric detectors would be required to detect different types of compounds. Consequently, the routine use of these types of detection devices for early and wide-spectrum warning is limited.

Other types of detectors are less selective and will respond to many chemicals. Flame ionization detectors respond to a large number of organic compounds without discrimination. Thus, they would not be suitable detection devices for CWAs or TICs in the field where the potential for exposure to specific toxins is unknown.

Photoionization detectors detect chemical compounds by ionizing them with UV light at certain electron volt energy levels. The only specificity of PID is that the substance with ionization potentials higher than the UV source will not be ionized and this will not cause a response signal. Similar to flame ionization detectors, photoionization detectors cannot distinguish among signals produced by different compounds. Their usefulness as CWA or TIC detectors in the field is limited because responses are derived from all substances that can be ionized, rather than only target compounds.

Technically speaking, any chemical that can be ionized by an ionization source will be detectable by the ion detectors. Some of these ion detectors can provide limited identification of diverse chemicals in an air sample. For instance, ion mobility spectroscopy devices, commonly referred to as IMS detectors, measure the time elapsed for the ionized species under atmospheric pressure to form the ion mobility spectrum. Thus, compounds can be identified through their respective characteristic mobility. However, because of the nondiscriminatory ionization process, IMS detectors are subject to potential interference by other compounds. It is necessary to provide means, such as limiting the number of peaks stored in the library or using more than one peak for substance identification, to improve the detectors identification capability and minimize false alarm potential.

Nonselective detectors are more suitable for field applications if only a broad-spectrum early-warning system is desired. A nonselective detector yields signals to various chemicals simultaneously, and thus could be useful in civilian arenas as an initial survey given that the chemicals used by terrorists in these scenarios are unpredictable. It would be necessary then to supplement the detection of certain compounds with more specific detection device(s) to identify potential CWAs and/or TICs.

It is important to recognize that there is no single detector that is absolutely selective or nonselective. A selective CWA detector may respond to other chemicals that possess similar properties. The number of chemicals a selective detector responds to is somewhat limited. Less-selective types of detectors can respond to more chemicals than selective detectors, but their responses cannot immediately be attributed to CWAs, TICs, or nontoxic substances. Therefore, a backup device with greater selectivity would still be required to assist in determining the precise nature of a given detected compound.

5.2 SENSITIVITY

Sensitivity of a detector or a method is a measure of its ability to discriminate between small differences in analyte concentration. The detector or method is more sensitive when a large change in signal intensity is observed for a small change in concentration. The instrument may also show a different degree of sensitivity toward low-concentration samples compared to high-concentration samples. Figure 5.1 is a generalized calibration curve observed for selected instruments. In the lower concentration range, the relative change of signal strength with the change in concentration (the slope of the curve) is greater than in the higher concentration range. Therefore, the detector is said to be more sensitive in the lower concentration range. When concentration increases out of the linear detection range of the instrument, the detector response starts to lose sensitivity and may reach a plateau (the output signal does not change as concentration increases).

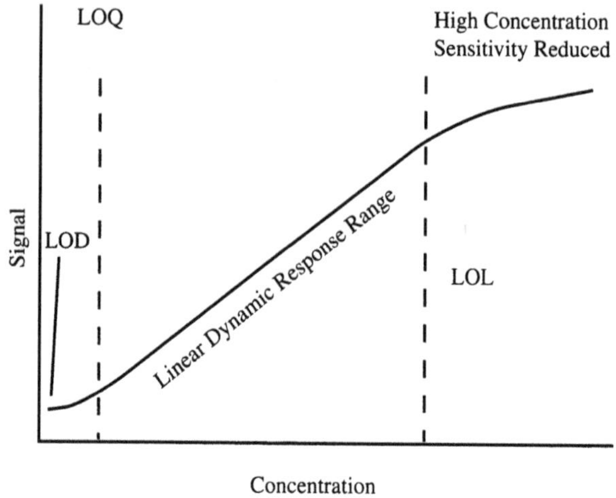

Figure 5.1 Response curve, limit of detection (LOD), limit of quantification (LOQ), and limit of linearity (LOL).

5.3 LIMIT OF DETECTION

A detector's limit of detection (LOD) is the lowest concentration level that it can identify with a certain degree of confidence. There are many definitions of LOD, such as the concentration at which the response signal generated is three times the instrument noise level. Here LOD is referred to as the minimum detection level (MDL) of concentration that will consistently cause the detector to alarm. It is affected by background noise and blank signals. The LOD of a detector may vary widely for different chemicals. Environmental and operational conditions could drastically affect the LOD. Manufacturers normally provide LOD information obtained under optimum conditions.

In general, detectors with lower LODs would be more suitable. As discussed in Chapter 4, the government has a set of exposure-level guidelines for numerous CWAs, TICs, and scenarios. A suitable detector should provide warning well before the IDLH level is reached to permit proper evacuation. Each type of operational scenario may require a different LOD.

5.4 RESPONSE DYNAMIC RANGE

After determining selectivity and sensitivity, dynamic range of the detection response is also a factor to be considered. Response dynamic range is the concentration range between the limit of quantification (LOQ) and the limit of linearity (LOL). The LOQ is the lowest concentration at which quantitative measurements can be made. The LOL represents the concentration level at which the calibration curve departs from linearity. LOD and LOQ are different in that the LOQ is usually somewhat higher than the LOD (Figure 5.1). Some contemporary detectors have large dynamic ranges of up to five or six orders of magnitude, while others have a dynamic range of only one or two orders of magnitude. When the concentration of a sample exceeds the LOL of the detector, the sensitivity of the detector is low. An increase in sample concentration may not be reflected in the observed or reported response. The smaller the dynamic range, the lower the sensitivity at lower concentration levels. The greater the dynamic range, the greater the sensitivity at lower concentration levels, and thus, the larger the concentration range in which a quantifiable response is generated.

For an analytical instrument, response dynamic range is crucial. However, for a field instrument, detecting the presence a toxic chemical in the air at a given concentration or higher is more important. Precise quantification of chemical composition is secondary provided that the detection level meets guidelines set by government agencies (for example) to identify hazardous substances.

5.5 QUANTITATIVE ANALYSIS CAPABILITY

Analysis performed using laboratory instruments can provide good quantitative results, because the laboratory environment is well controlled and calibration curves can be obtained with standard samples. Field instruments have to operate under harsh environmental conditions where temperature and humidity level can vary widely. Establishing standard calibration curves for an emergency situation is not practical. Field detectors can only be precalibrated in the laboratory for field applications, based on the assumption that the detector will perform the same or similarly under field conditions. Whether this assumption is valid depends on the relative ability of the detector to perform under field environmental conditions and stability of the detector after calibration. Many factors could affect the validity of this assumption. Environmental conditions such as humidity, dust, temperature, and potential interfering substances could affect quantitative or qualitative results.

Fortunately, in many situations precise quantitative analysis is not required. It is more important to know qualitatively and semiquantitatively whether the concentration of targeted chemicals surpasses certain levels. Therefore, semiquantitative analysis and qualitative analysis could satisfy such requirements. A semiquantitative detection result would indicate the approximate concentration range instead of an actual concentration level. Qualitative types of detectors only indicate whether targeted chemicals are detected at above-LOD levels, and do not display approximate concentrations.

5.6 FALSE ALARM RATE

False positive alarms and false negative alarms must be addressed when making decisions on appropriate detectors.

A false positive alarm occurs when the instrument reports that the sample contains targeted chemicals when in fact it does not. These alarms can be caused by various factors, depending on the specifics of techniques used. For example, phosphorus and sulfur compounds, such as pesticides, would generate a false positive

CWA alarm when a flame photometric detector is used for detection. This is because flame photometric detectors are made specifically to respond to chemicals containing phosphorus and sulfur. Detectors are always subject to yielding false positive alarms since no detector can be made 100% selective to only targeted chemicals. Most detection devices are made to detect multiple compounds. Therefore, a different technique may be necessary to serve as a backup to confirm alarms to reduce the potential for false positive alarms.

False negative alarms occur when instruments fail to respond to targeted chemicals that are present in a sample. These responses are viewed as more problematic than false positive alarms, because failure to produce a necessary alarm may lead to dangerous/disastrous situations. Causes of false negative alarms include changing environmental conditions; humidity effects; presence of interfering chemicals that mask normal detection capabilities; and detector malfunction, such as improper calibration and detection algorithm deviations.

5.7 RESPONSE TIME

Response time is the time required for the detector to respond to the targeted chemicals after an analysis cycle is started. The elapsed time for an alarm to occur after the detector is exposed to the targeted chemical at different concentration levels is another important factor. Response to higher-level concentrations must be fast and immediate, whereas somewhat slower response times for very low concentration levels may be acceptable.

Some detectors may require a certain amount of time before they can respond with an alarm. For example, detectors with a sample preconcentrator and/or a gas chromatography (GC) column in front of the sensor will require more time for analysis than detectors without such attachments. Response time is a function of vapor concentration for most direct-detection gross-level alarm devices.

Detectors that require thermal desorption, with or without GC column separation, will have a fixed response time. Therefore, regardless of the concentration level in the sample, response times for such devices will be approximately the same. The elapsed time necessary for analysis may be too long in some instances to permit warning in time for effective evacuation. For example, when a preconcentrator (which concentrates chemicals from vapor) is involved, response time will be that much longer than for a device without a preconcentrator.

Often, a GC column may be used to separate chemicals in a sample to provide specificity, and thus reduce the false alarm rate for nonselective detectors. Several minutes may be necessary for the process to complete. The cycle time depends on settings used for the GC elution. Consequently, detection by such instruments will require much longer response time than direct sampling types. Devices equipped with very low LODs with excessive response times may not be the best choices for a given operation because they lack the capacity for rapid warning in the event of high concentrations of the target compound. For postincident management scenarios, the more selective and sensitive types of instruments that include the use of sample concentration and GC separation would be useful.

In attempts to overcome the abovementioned risks, detector manufacturers have incorporated several functions to permit the choice between fast-response mode and high-sensitivity mode (preconcentration mode), or between the fast-response mode and the GC mode (less prone to false alarms).

5.8 RESISTANCE TO ENVIRONMENTAL CONDITIONS

Environmental conditions, including temperature, humidity level, dust concentration in the air, wind, and levels of contamination can affect the performance of a detector drastically. For military operations, temperatures in the field can range from –30°C to above 50°C with relative humidity levels ranging from less than 10% to 100%. It can also be very dusty, windy, and/or the air may contain diverse pollutants at varying levels. Ideally, a field detector must be operable under all environmental conditions, and should maintain its designated functions regardless of field conditions. However, all existing commercially available detectors are affected to one degree or another by environmental conditions. Operational temperature and humidity ranges are available from the manufacturers. Other factors that could affect their detection capabilities are not readily made available. Determining whether the detector can operate in the intended environment is crucial during the process of selecting detection devices.

5.9 SETUP AND WARMUP TIME

To respond to an emergency situation, it is desirable that a detector can be turned on and ready to operate within a short time. Setup time is the time needed to prepare the detector so that it can be powered up. Warm-up time is the time required for the detector to become ready for analysis after it has been turned on. For handheld detectors, the setup time is usually minimal because they are usually self-contained with only the necessary batteries as separate parts. Setup consists of inserting the battery pack. However, a period of warm-up time is normally required for any detector to reach operational status. This time could range from few seconds to half an hour or longer depending on the ambient temperature and relative operational parameters of the detector type. In general, the device goes through an internal self-check protocol to satisfy preset parameter requirements before it is ready for analysis functions.

5.10 CALIBRATION/VERIFICATION IN FIELD APPLICATIONS

Proper detector functioning will often require verification of capacity to perform upon each startup. For a chemical detector, the process is confirmed using known nontoxic chemicals as simulants of targeted compounds. Ideally, the proper operation of the detector can be verified with simple simulant checks that do not require complicated correlation calibration procedures before each use.

5.11 OTHER FACTORS

For a field detector, portability is another important factor to be considered. Portability, or whether the device can be transported, includes any support equipment required for operation. If transport and operation by a single person are considered important, then portability means a lightweight device. In situations where vehicle transport is acceptable or desirable, larger and heavier equipment (which in turn provide better detection and more refined analyses) carried by a vehicle are considered portable.

Field detectors must be transported from place to place by ground, rail, water, and air transport, and are subjected to extreme environmental conditions in storage. Rough handling in emergency situations is to be expected. Therefore, detector durability is another consideration.

In general, batteries are used to power handheld devices. Battery life is another logistical factor to be considered. Some detectors may require specially designed batteries, and thus finding replacements or recharging spent batteries in the field is difficult. Therefore, the chosen detector should ideally be operable through the use of two or more alternative power sources, and battery life must be sufficient to last through an entire mission. Fortunately, most modern handheld detectors are designed to permit the use of alternate power sources.

Operator training requirements are another factor to consider. Detector quality is only as good as an operator's understanding of instrument features and functions. Many modern instruments require a certain amount of special training to fully utilize their capabilities. Fortunately, most field detectors are designed to be user friendly; very little training is required to operate and maintain them.

However, a certain degree of data interpretation may be necessary to determine whether the data obtained from a given device is valid. Detailed knowledge of detector characteristics enables the operator to assess the likelihood of false positive response. Many field detectors do not directly display the concentration level of target chemicals that produce alarms. They may show bar graphs or other indicators on displays that show ranges of certain concentration levels. While knowing the exact concentration of the target agents may not be necessary, the ability to detect high concentration levels could prove to be vitally important. Operator training courses should emphasize result interpretation.

Operation costs should include equipment purchase, maintenance, and consumables costs. Cost comparisons should be based on the cost per analysis per chemical. For example, the purchase price of certain colorimetric detectors may seem relatively low but each analysis will incur additional consumables costs. Thus, the final cost of an operation could be relatively high compared to automatic continuous types of devices due to the cost of required consumables. Thus, although an automatic detector may cost tens of thousands of dollars initially, it does not require large amounts of consumables, it detects a number of chemicals at once, and it has a very long operational life.

There are other secondary concerns, such as the time required to clean the residual from the previous sample, waste generation, storage effects, maintenance frequency, whether the device can be easily decontaminated, and whether data can be saved for later analysis. The most important consideration is that the chosen detector must be able to provide the necessary reliability, sensitivity, and sufficient selectivity for a given operation scenario.

The perfect detector does not exist and no single detector possesses all the desirable features and functions. By weighing the factors mentioned above together with the operation scenario, it may be possible to find a detector that approaches meeting all or most requirements.

CHAPTER 6

Ion Mobility Spectrometry

Many of today's CWA detection devices are based on ion mobility spectrometry (IMS), which separates ions based on their different drift velocities in an electric field. Besides being used to detect CWAs, IMS has also been widely used to detect explosives and illicit drugs. Recently, its applications have been expanded to many other areas, including the detection of toxic industrial materials (TIMs).

Theoretically, the time for an ion to drift through an electric field within the drift region under ambient atmospheric condition depends mainly on its shape, mass, and charge. When the strength and length of the electric field are kept constant, the spectrum of an ion's mobility through the drift region is generally referred to as its "signature." A specific ion will have a specific spectrum under specific conditions. Mobility spectra of various ions obtained under controlled conditions are catalogued in libraries residing within the detector. These libraries are used to enable identification of an unknown sample through spectrum matching. Mobility (K) is defined as

$$K\left(\mathrm{cm}^2/(\mathrm{V}\cdot\mathrm{s})\right) = \frac{\mathrm{v}(\mathrm{cm/s})}{\mathrm{E}(\mathrm{V/cm})} \tag{6.1}$$

where v is the drift velocity of an ion and E is the electric field gradient.

IMS technology has many advantages: rapid analysis, high sensitivity, and low limit of detection (LOD). It does not require any sample preparation in most circumstances, particularly gas samples. For vapor samples, the LOD for most CWAs is at parts-per-billion (ppb) to low parts-per-million (ppm) levels within a few seconds. For solid or liquid samples, as little as a few nanograms (10^{-9} g) of a chemical can be detected.

IMS is a nonselective technology, but with high identification power due to its ability to separate different ions in the drift tube. This is a great advantage over other nonselective techniques. Thus, it can detect a wide range of chemicals and identify them. Most IMS instruments are made to operate under atmospheric conditions.

They do not require the use of additional carrier gases, vacuum pumps, or specialized power supplies. Most field-portable instruments use a weak radioactive ionization source to provide the ionization energy. These features significantly simplify the instrumentation and permit detector miniaturization for field operations. This technique has received a great deal of attention and its application is being rapidly expanded.

6.1 PRINCIPLE OF OPERATION

Figure 6.1 shows a diagram showing the operational principle of IMS detection. The process includes sample introduction, ionization, ionic drift, collision and diffusion, ion collection, and signal generation. Sample vapor is drawn into the detector via the inlet into the ionization region. Many detectors use a membrane barrier to limit the entry of moisture into the drift cell. Some detectors eliminated such membrane to achieve faster responses. Molecules in the vapor sample become ionized by the ionization source to form positive and negative ions according to the physical and chemical properties of the individual molecule. This ionization process occurs under atmospheric conditions. Various ionization sources may be used: radioactive, UV, electron spray, or corona discharge. The use of a radioactive source is by far the simplest and most common. Either positively or negatively charged ions enter the drift tube with an electrical field gradient within a very short time gap (microseconds). While in the drift region, ions of different shapes and sizes formed from a variety of chemicals contained in the vapor sample undergo diffusion and collision with other particles while drifting along the electric field gradient.

6.1.1 Drift

Ions move along the drift tube forced by the electric field gradient applied to the tube. Depending on the strength of the electric field and the shapes and sizes of the ions, the ions will reach the ion collector at other end of the tube at different, but all very short (milliseconds), times.

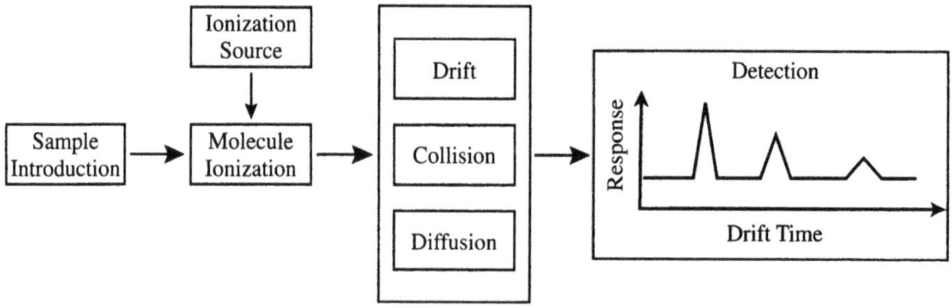

Figure 6.1 Block diagram of ion mobility spectrometer analysis procedure.

Theoretically, force (F, Newton) exerted on the ions resulting from the electrical field (E, volt) applied to a charged particle with a charge of q (Coulomb) is described as

$$F = q \times E \tag{6.2}$$

This force accelerates the ions to move along the electric field gradient toward the ion collector. Positive ions are forced to move from a higher to lower voltage and negative ions move from a lower to higher voltage end of the gradient field. The drift times of ions should be in microseconds for a 10-cm-long electric field with a gradient of 200 V/cm (a typical configuration of drift tubes) when under vacuum. Because the drift tube for IMS detectors is not being operated under vacuum, the observed drift time is actually on the order of milliseconds.

6.1.2 Collision

A drift flow, usually a dried airflow, is normally added in the opposite direction of the ion drift. Since the drift tube is not under vacuum, the ions in the drift tube collide with other molecules in the drift flow. The ions are slowed or stopped by these collisions and then accelerated again by the electric field gradient. This drift-collision-drift process occurs throughout drift time until the ions finally reach the collector. Overall drift time is ultimately controlled by the collision frequency when the electric field gradient is held constant. Ions with larger effective collision cross-sections collide with molecules in the drift flow more frequently than ions with smaller cross-sections. Consequently, overall drift velocity is slower for larger ions. Therefore, larger ions have longer drift times while smaller ions reach the collector earlier. This collision process separates the ions having different collision cross-sections and determines the mobility of the different ions.

6.1.3 Diffusion

In the drift tube, ions also disperse in the tube through diffusion. Diffusion is a concentration gradient–driven process. This diffusion process depends on the nature of the ions and the drift tube temperature that produce the concentration gradient of the ions as they migrate within the tube. The concentration gradient tends to broaden the detection peaks for respective ions in an ion mobility spectrum. This diffusion process is one of the reasons why the longer drift-time ion peaks are broader.

6.1.4 Detection

When the ions eventually reach the ion collector, they lose their acquired charges to the collector, and an electrical current is generated. This electrical signal is being processed to generate the spectrum or other desired formats using a microprocessor. The substance will be reported as detected when the generated information matches the criteria set and stored in the detector's libraries. An audible and/or visible alarm is generated based on the logic used by the instrumentation.

6.2 INSTRUMENTATION OF TYPICAL IMS DETECTOR

A typical configuration for an IMS detector is illustrated in Figure 6.2. The instrument includes sample inlet, ionization region, ion injection gate, drift region, ion collector, and microprocessor.

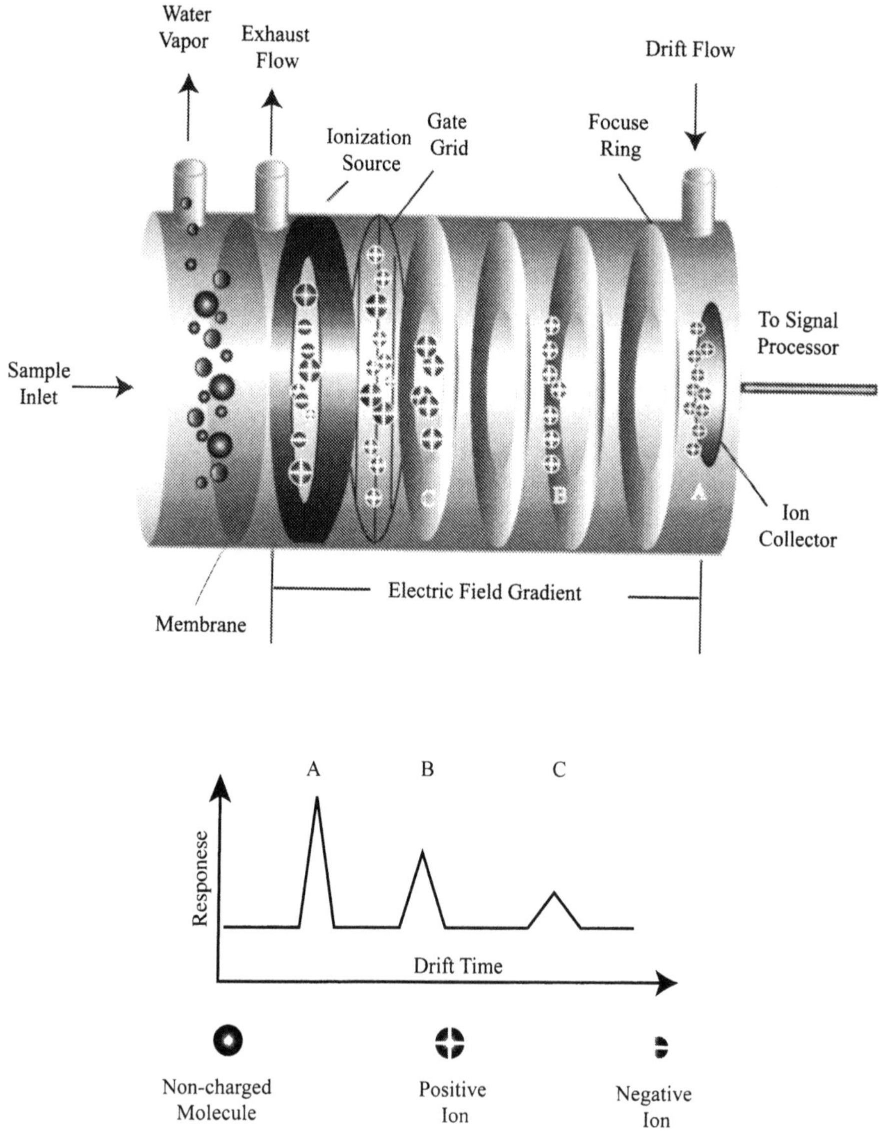

Figure 6.2 Schematic diagram of IMS.

6.2.1 Sample Inlet

The sample to be analyzed enters the detection device through the sample inlet. Its design depends on the type of substances monitored. IMS instruments have been used successfully to analyze CWAs, particulate and vapor samples for illicit drugs, and explosive chemicals. A wiper is used to collect particulate samples by sweeping the surfaces of an object. The swiped sample is heated so that the substances adhering to the swap are vaporized and drawn into the IMS cell for analysis. Similar sampling techniques may also be used for detecting CWA and TIC residues on surfaces. The technique is especially useful for detection of persistent chemicals that have very low vapor pressure. As discussed earlier, these low volatility chemicals are more persistent, and higher concentrations can be found on surfaces where concentrations in the air may be too low for efficient detection.

Most CWAs and high-hazard TICs have relatively high vapor pressure, however. Air contaminated with any of these compounds constitutes a major health risk. Detection of the highly volatile toxic compounds must be done as quickly as possible. The only way this is possible is by direct analysis of the air sample. IMS detectors are designed for this purpose. These detectors can produce very fast responses once the sample reaches the IMS cell. The unknown vapor sample is drawn into these detectors directly without requiring sample preparation. Inlet materials are selected to permit efficient sample transfer. The sample path is generally heated for added sample transfer efficiency.

Moisture content in the sample presents a significant interference problem and reduces IMS detector's identification capability. Commonly, a thin heated membrane that has a low permeation rate for water, such as a dimethylsilicone membrane, is added in front of the IMS ionization region. This membrane is used to limit the amount of moisture that could enter the ionization region to counter the moisture effect while permitting the compounds of interest to permeate. Chemicals in the sample will diffuse through the membrane into the ionization area while most of the water molecules together with other chemicals that also have low permeation rates are carried away by the carrier flow without entering the ionization region. The membrane also serves as a barrier to block dust or other particulate from entering the drift tube.

The use of a membrane, while it reduces the moisture effect, has some disadvantages. The targeted sample must pass through the membrane via the diffusion process before analysis is possible. Diffusion or permeation is a relatively slow process that requires time to reach equilibrium. Therefore, the time to reach the maximum response is longer compared to detectors that do not use a membrane. The membrane can easily become contaminated with dust and other low volatility substances. When that happens, it will require either a long time to clear or maintenance action to replace the membrane. The membrane is very delicate and easily broken. Broken membranes cause the detector to malfunction. Membranes also prevent some of the targeted chemicals from entering the ionization region, which in turn affects detection sensitivity.

6.2.2 Ionization Region

6.2.2.1 Ionization Sources

After diffusing through the membrane, sample molecules enter the ionization region of the detector and become ionized by the ionization source. In most IMS detectors, especially handheld ones, a weak radiation source, $_{28}Ni^{63}$, is used to provide the ionization energy. $_{28}Ni^{63}$, a radioactive isotope of nickel, has a half-life of 100.1 years (National Nuclear Data Center, www.nndc.bnl.gov) and low decay energy of 0.067 meV. $_{95}Am^{241}$ is also used in some detectors as the ionization source. The alpha-decay isotope of americium ($_{95}Am^{241}$) has a half-life of 458 years and decay energy of 5.640 meV. Unlike the flame ionization process, ionization of compounds by $_{28}Ni^{63}$or $_{95}Am^{241}$ does not break the molecules into fragments. Ionization using a radioactive source reduces the signal noise and power consumption of a detector since it does not require electrical power for the ionization process. This greatly simplifies the instrumentation of IMS detectors. Reduced power consumption provides advantages for handheld detectors that must be operated using battery power.

Other ionization sources, such as "corona discharge" and "electron spray" are also used in some IMS detectors. Due to respective limitations compared to the use of a radioactive ionization source, they have not been as widely used in commercialized handheld detectors as those that use a radioactive source. On the other hand, the use of a radioactive ionization source is limited by strict regulatory controls associated with radioactive materials.

$_{28}Ni^{63}$ is a common radioactive source used in many CWA detectors. We discuss its application in more detail as an example to explain the detection process of an IMS device.

6.2.2.2 Ionization Processes

As $_{28}Ni^{63}$ decays, electrons (e^-) are released to become $_{29}Cu^{63}$ through beta decay:

$$_{28}Ni^{63} \rightarrow {}_{29}Cu^{63} + e^- \quad (6.3)$$

Electrons released by decay react with molecules in the air to generate positive and negative ions that accumulate in the ionization region to form a pool of ions. The possible types of ions in this pool include $H^+(H_2O)_n$ (n = 1 to 10), NH_4^+, NO^+ as positive ions, and $O_2^-(H_2O)_nX_m$ (X represents a neutral vapor molecule) as the predominant negative ions. Some detectors use additional reactant gas in the ionization region to increase the ionization efficiency of compounds and alter potential interference effects. By transferring associated charges of the ions through different mechanisms such as proton transfer and electron attachment to other molecules, molecules of the targeted chemicals are ionized to form the ions that are detected in the ionization region. An example of proton transfer for the formation of positive ions is

$$M + H^+(H_2O) \rightarrow MH^+ + H_2O \quad (6.4)$$

where M is the molecule of the targeted chemical that reacts with a protonated water molecule to form a positive ion for detection and identification.

There are several ways to form negative ions. Ions can be formed by electron attachment as

$$e^- + M \rightarrow M^- \tag{6.5}$$

Electron transfer is another way that negative ions are formed:

$$O_2^- + M \rightarrow M^- + O_2 \tag{6.6}$$

Negative ions can also be formed through proton abstraction reactions:

$$M + O_2 \rightarrow (M - H)^- + O_2H^+ \tag{6.7}$$

6.2.2.3 Charge Competition

The strength of the radioactive ionization source activity used in a detector is regulated for safety and environmental concerns. Since the number of ions that can be produced in a given unit of time is related to the activity of the ionization source, limiting the activity of the ionization source limits ion production. Consequently, the number of sample molecules in the ionization region, especially at high concentration levels, may exceed ionization capacity. As a result, not all molecules in the ionization region will be equally ionized. The molecules have to compete for charges to become ionized. In the formation of negative ions, molecules with high electron affinity are ionized more easily than those with low electron affinity. Electron affinity is the energy released from an atom when an electron is added. For example, oxygen is a major component of air. Thus, oxygen exists as a background peak in the negative mode. When a sample containing chlorine gas enters the ionization region, the chlorine in the sample competes with oxygen for the negative charges. Because chlorine has a higher electron affinity (2.5 eV) than oxygen (0.45 eV), it is more easily ionized. The response to oxygen therefore becomes smaller as the peak for chlorine appears.

An advantage associated with this charge competition is that it provides an avenue toward the elimination of possible interference from nontargeted compounds in the vapor sample. However, when several compounds with similar charge affinity exist together with the CWA or TIC, they are ionized proportionately to their respective concentrations in the vapor sample. The greater a compound's concentration in the sample, the greater the detector's response. A problem arises when the CWA or TIC has lower charge affinity than interference compounds in the sample, as the interference compounds are more easily ionized than the target compound(s). Relative abundance of interference substances could consume the available ionization energy, affecting the IMS spectrum that in turn may result in false alarms.

6.2.3 Ion Injection Gate

The ions generated in the ionization region are drawn into the drift region by the electric field gradient applied to the drift tube. The generated ions are not allowed to continuously enter the drift region. If the ions continuously enter the drift region, the technique would not be useful. The ions are "injected" into the drift region in a very short pulse controlled by an electronic gate or shutter. Basically, the electronic gate is another electrical field that is vertical to the electric field of the drift tube produced by grids. When set to "open," the field of the shutter is removed and ions enter the drift region. This "open" time is kept as short as possible to achieve good separation of different ions. When the shutter is set to "close," the electric field generated by the shutter prevents additional ions from entering the drift region. This "open-close" process is repeated for each analysis cycle.

Not all ions are injected into the drift tube when the gate is open. The electrical field gradient between the ionization region and ion collector controls the type of ions entering the drift tube. Either positive or negative ions are injected into the drift tube according to the electric field gradient of the drift tube. When the ion collector side has higher voltage than the ionization side (negative mode), negative ions are injected into the drift tube by the electric field. Positive ions enter the tube when the voltage of the collector side is lower than the ionization region (positive mode). Figure 6.2 shows the positive mode. The negative mode works in a similar fashion, except that negative ions rather than positive ions are injected into the drift region by reversing the electric field. The ions that could not enter the drift tube together with the compounds in the vapor that were not ionized are swept out of the detector by the drift flow.

6.2.4 Drift Tube

The drift tube is composed of rings of different voltages to create a homogeneous electric field gradient throughout the entire length of the drift tube. The tube is usually several centimeters long for handheld detectors. Benchtop instruments and some instruments built for research purposes may use a longer drift tube to increase peak resolution capability. Tube diameter can vary from a few millimeters to a few centimeters. The voltage applied to the drift tube is on the order of several hundreds of volts per centimeter. Within the ionization region, both positive and negative ions are generated. The electrical field gradient of the drift tube can be alternated between positive and negative mode to permit detection of both positive and negative ions.

Within the drift tube, a dry airflow called drift flow is used to remove neutralized ions from the collector along with ions that were not injected into the drift tube from the detector. Given that the voltage applied to the drift tube remains constant, the variation of temperature, pressure, and moisture level in the drift flow will significantly affect ion mobility and cause drift time shifts. When the moisture level in the drift flow is high, ion clusters surrounded by water molecules are possible. This reaction changes the shape and size of targeted ions. Consequently, the mobility of ions clustered with the water will differ from the ions without water molecules. Therefore, for reliable detection, it is very important to keep the air supplied to the drift flow dry and free from other organic matter to maintain consistent mobility of the ions of interest.

For instruments without adequate control of the drift tube temperature, change in the ambient temperature becomes an important factor that affects the detection algorithm. Such effects must be compensated for. Fortunately, the effect of this factor can be estimated and compensated for during the signal processes. Alternatively, this effect can be overcome by providing adequate temperature control of the drift tube. Pressure changes also cause a similar effect to that of temperature changes because both temperature and pressure affect density of the gas within the drift tube. The pressure-induced effect can also be estimated and compensated for during signal processing.

6.2.5 Ion Collector and Signal Processor

Various ions collide with the ion collector located at the end of the drift tube at different times due to the separation process that occurs within the tube. These ions release their acquired charges and return to their neutral states. The time interval between when the ions are injected into the drift region and when the ions collide with the collector is recorded as the "drift time." These drift time data are gathered and used to calculate ion mobility together with correction factors to compensate for temperature and pressure effects. An electric current or voltage is generated and processed by the signal processor into a series of peaks representing the relative drift times for the various substances. The signal strength is proportional to the number of the ions that collided with the collector electrode. Therefore, the amplitude of the peak generated can be used as an indication of relative concentration of the detected chemicals.

Signal processing correlates the relative drift times and signal characteristics of the sample and compares the results to the detector's internal libraries for pattern matches. If the information generated from the detection signal closely resembles the information stored for a target substance, the detector reports that the substance has been detected and generates an alarm. Because of the limited separation capacity in the short drift tube, the number of targeted compounds programmable for detection must be limited to avoid peak overlapping that leads to detection interference and frequent false alarms.

6.3 TECHNIQUE SPECIFICATION

6.3.1 Detectable Substances

Theoretically, any molecules that can be ionized are detectable by IMS-based detectors. Researchers have shown that IMS detectors can be used in various applications, such as detecting explosives, illicit drugs, CWAs, and many TICs. It has also found other applications. At present, interest is greatest in instruments that can detect multiple chemicals to minimize the number of devices necessary for an operation. This focus has made IMS detectors desirable because of their nonselective but high identification power attributes.

6.3.2 Selectivity

IMS-based detectors are nonselective, they will detect many different vapors, including those that are not of primary interest. Because of the drift tube's ability to separate chemicals, IMS instruments can identify multiple chemicals in the sample. When these detectors are used in the field, sample matrices are not as simple as in the laboratory. Field samples can be quite complex and may contain many different compounds. Unfortunately, many false alarms/identifications can occur because of limitations on resolution. False positive alarms can become problematic when mobility of nontargeted and targeted chemicals is similar. The larger the number of chemicals that the detector can identify, the greater the number of assigned detection peaks. When these peaks are too close together, any substance may have a drift time similar to one of the assigned peaks and a false positive alarm occurs. So, the number of substances included must be limited to minimize the potential of frequent false alarms.

There are several ways to reduce the false positive alarm rate. Adding a gas chromatographic (GC) column in front of the detector may help to reduce the false alarm rate by further increasing the detector's chemical separation power. The additional hardware will result in longer analysis time and more complex instrumentation. Some IMS detectors use chemical additives to reduce false alarms. The addition of a reactant chemical may (1) form a different peak that can be segregated from the peak generated by the target chemical, or (2) suppress the detector response to certain potential interfering chemicals of lower charge affinity, thereby increasing selectivity. Only molecules with charge affinity higher than that of the additives will be ionized and detected.

The idea is that it may be possible to reduce the false positive alarm rate by using more than one peak to identify a target chemical. The probability of two chemicals generating a common peak is much greater than that of the two chemicals generating two or more common peaks at the same time. Therefore, using more than one peak for identification could drastically reduce the false positive alarm rate.

False negative alarms can be an issue when a complex sample matrix is analyzed. As discussed above, molecules with high charge affinity are more easily ionized. If the sample contains chemicals that are ionized more easily than the targeted chemicals, the response to the targeted components may be suppressed. This will result in a false negative response. It is possible to reduce the false negative effects by choosing a reactant additive that could increase the ionization propensity of the targeted chemicals. False negative responses must be addressed. More aggressive research to address this issue is warranted.

6.3.3 Sensitivity and LOD

IMS detectors have reasonable sensitivity for low concentration samples. As shown in Figure 6.3, when concentration in a sample reaches a certain level, these detectors become insensitive to higher concentrations. This is due to the limited supply of ions in the ionization pool, which is insufficient for ionization of all molecules entering the ionization region. In general, the LOD of IMS detectors is

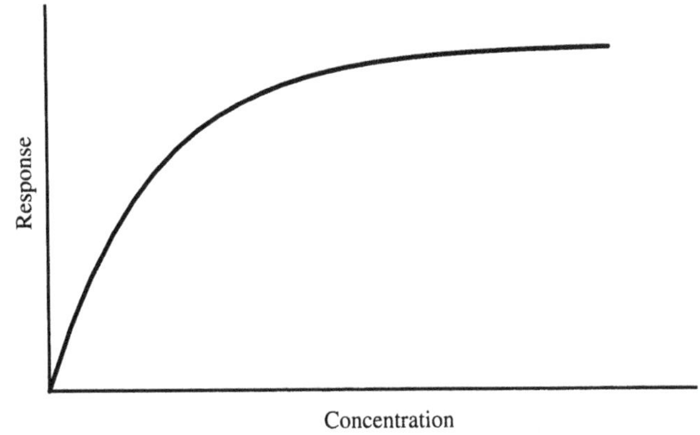

Figure 6.3 IMS response curve.

in the high parts-per-billion to low parts-per-million level, depending on specific chemicals and detector design. IMS detectors tested in the Domestic Preparedness Program showed that the LOD varied from detector to detector, but they are capable of detecting IDLH or JSOR concentration levels.

6.3.4 Response Dynamic Range and Quantitative Capacity

The dynamic response range of an IMS-based detector is less than two to three orders of magnitude, compared to five or six of other detection technologies. The concentration versus response relationship is exponential rather than linear. As discussed earlier, due to the limitation of total available ions in the ionization region, the response may reach its saturation level rather quickly such that a further increase in concentration will not lead to a much stronger signal (Figure 6.3), and thus the relationship is not linear.

However, the estimation of concentration at the low response range based on a linear relationship may be acceptable. For field instruments, the primary objective is to detect whether the sample contains the targeted chemicals. Therefore, the IMS detector's nonlinear relationship between concentration and response is not a great disadvantage for field applications.

6.3.5 Resistance to Environmental Conditions

Temperature, pressure, and humidity may affect the performance of an IMS detector. Potential problems associated with sample contamination may influence detection capability. Dust can be filtered with a particle filter. As discussed above, temperature affects ion mobility. Therefore, variation in environmental temperature affects the detection capability of the instrument. Some detectors operate with a set drift tube temperature. If and when the environmental temperature is higher than the set point, ion mobility changes, resulting in peak position shifts. The shifts can generally be predictable and could be compensated for during signal processing.

Some detectors are equipped with controlled higher tube temperature. The tube temperature can be set much higher than the environmental temperature to avoid shifts in peak positions caused by temperature variance.

Changes in atmospheric pressure will affect ion drift time. This effect has been well studied and can be easily compensated for.

Humidity effects are more complex. The humidity level of the sample cannot be controlled in field environments and may vary from very dry (close to 0% in the desert or in winter) to saturation level (rainy weather). The humidity level affects the species entering the ionization region. Various chemicals can be hydrated to form water clusters and thus different ions. Therefore, peaks observed under high humidity may shift away from the position (mobility) observed at low humidity. When the shift is significant, the detector may fail to identify a targeted chemical. Using a membrane may help to reduce this effect. Testing targeted chemicals at various levels of relative humidity is necessary to fully understand their effects.

A drift flow is used to keep the drift tube clean. It is assumed that there is no chemical reaction occurring in the drift region, and that the ions are in the same form as in the ionization region. When the drift flow is wet, density of the drift flow may change, which causes changes in collision frequency and ion mobility. Wet drift may also react with the ions in the drift tube and change their mobility. Both effects may cause false negative or false positive alarms. Fortunately, the humidity level of the drift flow is controllable through the use of desiccant. Effects can be reduced to a minimum if the humidity level of the drift flow is closely monitored and dried as necessary.

6.3.6 Other Specifications

IMS detectors are mostly designed as direct analysis devices without sample preparation. Response time is seconds if preconcentration and/or GC is not involved. Response time will be contingent on the preconcentration time and/or GC time, if they are used.

Setup time is minimal for a handheld IMS detector. It involves inserting the battery pack for power. Detector warmup may run from a few minutes to half an hour. Most current detectors have internal function and parameters checking procedures installed. This self-test feature verifies that proper operational parameters are satisfied before the unit is ready for use. After the warmup, detector calibration can be easily verified with simulants. If verification is successful, the device is ready to detect.

Handheld IMS detectors can be very small, weighing less than 20 lb, and can operate on batteries. Operation and maintenance of IMS detectors are quite simple, requiring only minimal training. Most detectors are designed to be user-friendly. Minimal or no data interpretation by the operator is required because the microprocessor performs most of the interpretation. The detector will show target substances found in the samples, and will trigger an alarm to alert the operator. The detector can be connected to a computer for more detailed data collection, if necessary.

The most important maintenance effort for IMS detectors is to maintain a dry drift flow. This effort requires routine checking and replacement of dryer cartridges.

6.4 APPLICATIONS

CWA detection using ion mobility technology has been developed for decades. The first CWA detector developed and employed by the U.S. military for field uses, the electrochemical M43® alarm, was replaced by the M43A1® alarm, which is a type of ionization detector. The M43A1 is called the Army Ionization Detector. The initial M43 detector was a wet chemistry type that could detect G-nerve agents and VX, but not HD. The M43 detector requires the use of an oxime solution circulating through an electrochemical cell for detection.

The development of the M43A1 eliminated the need for solution preparation. This device also provided faster and more reproducible responses to less than the IDLH level concentration of nerve agents. The M43A1, however, does not have HD detection capability.

During the mid-1980s, HD detection capability was added to the M43A1 with development of the Mustard Upgrade. Developers experimented with several different techniques, including the use of surface acoustic wave (SAW) sensors and adding negative-mode ionization detection. The effort led to development of a handheld chemical agent monitor called the CAM®. CAM detects the nerve agents in its G-mode or HD in its H-mode, but cannot detect both at the same time. The CAM has undergone several revisions since its inception, as experience gained led to more reliable detection and reducing false detection rates that are inherent to the ionization process. The improved versions include the CAM-2® (Figure 6.4), ICAM®, APD2000®, RAID-M®, IMS-2000®, ACADA-GID3® (Figure 6.5), Sabre2000® (Figure 6.6), and many others. Fact sheets on these devices are included at the end of the chapter, which show that all use IMS technology, but they were designed for different applications.

It is not possible to include discussion of all available detectors in this book. In subsequent chapters, as in the current one, examples are provided for specified technologies. Detectors mentioned in the text and listed in tables are not necessarily better or worse than others; they are presented as examples only.

The Lightweight Chemical Detector, or LCD® series (Figure 6.7), because it was developed more recently, differs slightly from the aforementioned detectors. The LCD uses the open-cell technique that does not contain a radioactive source, and

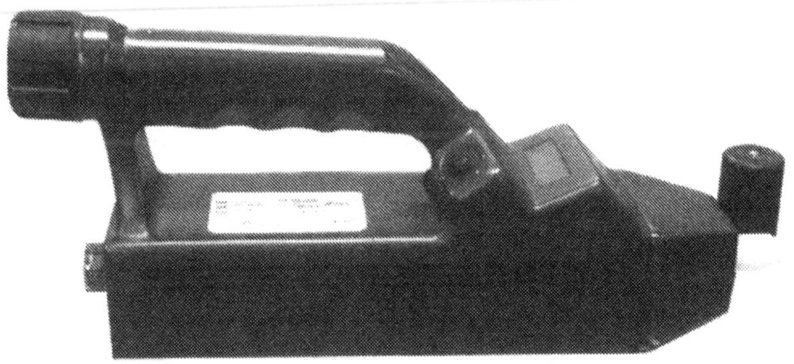

Figure 6.4 CAM-2 Detector. Photo courtesy of K. Y. Ong.

Figure 6.5 GID-3 Detector. Photo courtesy of K. Y. Ong.

Figure 6.6 Sabre2000 Detector. Photo courtesy of K. Y. Ong.

has eliminated the use of a membrane to protect the drift tube from moisture effects. Elimination of the membrane stemmed from the type of IMS detection used by the M90® detectors developed by Environics Oy of Finland. Environics applied technology similar to the M90, and developed the ChemPro 100®. Both the ChemPro 100 and LCD are currently involved in testing in hopes of replacing or supplementing the Joint Chemical Agent Detector (JCAD) should the current SAW technology fail to deliver.

From the above, it is clear that IMS has been the major technique used for CWA detection for the past 20 years. Until there is a better form of mature detection technology, IMS will continue to rank at the top of the list now and into the near future as the most accepted technology for CWA detection. It is especially true that due to the nonspecific ionization process and identification capability, TIC compounds can easily be detectable using this technique as well. Developing TIC detectors is a top priority for the Department of Homeland Defense in the war against terrorism since the September 11, 2001, incidents.

Figure 6.7 LCD Detector. Photo courtesy of K. Y. Ong.

6.5 FACT SHEETS FOR SELECTED IMS-BASED DETECTORS

Detector Name

ACADA GID-3 Automatic Chemical Agent Detection Alarm, M-22

Manufacturer/Distributor and Contact Information

Smiths Detection
2202 Lakeside Blvd.
Edgewood, MD 21040
Tel: 410-510-9100

Technique Description

This device is a portable CWA detector using IMS.

Chemical Detection Capability and Performance

The ACADA can detect nerve, blister, blood, and choking agents at the JSOR concentration level with near real-time detection. Operable at JSOR temperature extremes. Is type classified by the U.S. as the official CWA detector and alarm that has undergone extensive testing according to the military standards.

Other Features

Audio and visual alarm indicators. Can be upgraded to detect new compounds. Consumables include occasional filter and desiccant cartridge replacements. Meets military specifications, and is currently used in the field by the U.S. military.

Detector Name

APD-2000

Manufacturer/Distributor and Contact Information

Smiths Detection
2202 Lakeside Blvd.
Edgewood, MD 21040
Tel: 410-510-9100

Technique Description

This instrument is a handheld CWA and riot control agent detector that uses IMS. It is capable of providing agent identification, and has selectable operating modes including radiation detection capability.

Chemical Detection Capability and Performance

It can detect nerve, blister, blood, and choking agents at the JSOR concentration level with near real-time detection. It can also detect riot control agents such as the pepper spray and mace. Operable at JSOR temperature extremes.

Other Features

Audio and visual alarm indicators. Can be upgraded to detect new compounds. Consumables include occasional desiccant cartridge changes. Suitable to serve as a monitor in a CWA or civil defense scenario.

Domestic Preparedness Program Report

Domestic Preparedness Program: Testing of APD2000 Chemical Warfare Agent Detector Against Chemical Warfare Agents Summary Report, August 2000 (www2.sbccom.army.mil/hld/ip/apd2000_detector_download.htm).

Detector Name

CAM/ICAM/CAM-2 Chemical Agent Monitor

Manufacturer/Distributor and Contact Information

Smiths Detection
2202 Lakeside Blvd.
Edgewood, MD 21040
Tel: 410-510-9100

Technique Description

This instrument is a handheld, portable CWA monitoring detector that uses IMS. The CAM-2 has the added detection algorithm for identifying hydrogen cyanide, phosgene, and chlorine, in addition to conventional CWAs.

Chemical Detection Capability and Performance

It can detect nerve, blister, blood, and choking agents at the JSOR concentration level with near real-time detection. Operable at JSOR temperature requirements.

Other Features

Visual bar response indicators. Mode selection required. Can be upgraded to detect new compounds. Consumables include occasional filter and desiccant cartridge changes. Meets military specifications and is currently used in the field by the U.S. military.

Domestic Preparedness Test Report

Domestic Preparedness Program: Testing of the CAM-Chemical Agent Monitor (Type L) Against Chemical Warfare Agents Summary Report, August 2001 (www2.sbccom.army.mil/hld/ip/cam_typel_download.htm).

Detector Name

ChemPro 100 Chemical Warfare Agent Detector

Manufacturer/Distributor and Contact Information

Environics USA
4420 Eastport Parkway
Port Orange, FL 32127
Tel: 386-304-5252

Technique Description

The ChemPro 100 detector is a small, handheld CWA detector that uses Environics' patented "Open Loop Ion Mobility Spectrometry" that can operate IMS without the need for desiccant. The ChemPro 100 uses the Ion Mobility Cell™ for better sensitivity and selectivity.

Chemical Detection Capability and Performance

It detects nerve, blister, blood, and choking agents at the JSOR concentration level with near real-time detection. Operable at JSOR temperature extremes.

Other Features

Audio and visual alarm indicators. Can be upgraded to detect new compounds through changeable data libraries. Capable of storing 50 libraries of 60 gas classes each. No consumables except for occasional filter changes. Can be used as a CWA or TIC detector. Has data logging capability and is capable of networking.

Detector Name

IMS 2000 Handheld Chemical Warfare Agent Detector and RAID-M

Manufacturer/Distributor and Contact Information

Bruker Daltonics Inc.
40 Manning Park
Billerica, MA 01821
Tel: 978-663-3600

Technique Description

This detector is a handheld ion mobility spectrometer using ammonia as dopant to permit operation under an ammonia background.

Chemical Detection Capability and Performance

The detector identifies nerve, blister, and blood agents. Sensitivity is expected to meet the current JSOR requirements.

Other Features

Audio and visual alarm indicators. Interface for connecting to computer networks available. Free programmable libraries and software can be downloaded. Back-flush protection to prevent oversaturation.

Detector Name

IPDS Improved Point Detection System

Manufacturer/Distributor and Contact Information

Tradeways Ltd.
184 Duke of Gloucester Street
Annapolis, MD 21401
Tel: 410-295-0821

Technique Description

This device is a shipboard stationary IMS CWA detector that uses elongated ion mobility cells. It is capable of providing superior resolution to counter false alarms. The instrument has algorithm libraries to enable low-level detection of CWAs while filtering out common shipboard types of interference.

Chemical Detection Capability and Performance

It can detect nerve, blister, blood, and choking agents approximately at the JSOR/IDLH concentration levels within 1 min.

Other Features

Audio and visual alarm indicators. Can be upgraded to detect new compounds. Operable in a ship environment between 0° to 50°C, and meets military specifications.

Detector Name

Lightweight Chemical Agent Detector (LCD)

Manufacturer/Distributor and Contact Information

Smiths Detection
2202 Lakeside Blvd.
Edgewood, MD 21040
Tel: 410-510-9100

Technique Description

This instrument is a handheld CWA monitoring device using an advanced form of IMS that does not require a radioactive source for ionization. The unit is equipped with an algorithm for detection of hydrogen cyanide, phosgene, and chlorine, in addition to conventional nerve and blister agents.

Chemical Detection Capability and Performance

It can detect nerve, blister, blood, and choking agents at the JSOR concentration level with near real-time detection. The unit includes a detailed user interface that permits identification of detection class and level, as well as data logging and networking.

Other Features

Visual and audio response indicators. Mode selection. Can be upgraded to detect new compounds. Consumables include occasional filter and desiccant cartridge changes.

Detector Name

M-90A1-C Chemical Warfare Agent Detector

Manufacturer/Distributor and Contact Information

Environics USA
4420 Eastport Parkway
Port Orange, FL 32127
Tel: 386-304-5252

Technique Description

This device is a handheld CWA detector using Environics' patented "open loop ion mobility spectrometry" that can operate IMS without the need for desiccant. Unlike conventional IMS detectors, in that M90 the sample does not need to penetrate a membrane before reaching the IMS cell, which makes faster responses possible. It also includes a semiconductor cell for other detection purposes. Chemical identification is based on signal pattern recognition.

Chemical Detection Capability and Performance

It can detect nerve, blister, blood, and choking agents at the JSOR concentration level with near real-time detection. Operable at JSOR temperature extremes.

Other Features

Audio and visual alarm indicators. Can be upgraded to detect new compounds through changeable data libraries. No consumables except for occasional filter changes. Meets military specifications and is currently used by the armed forces of many nations. Can be networked and connected to a computer to update libraries and implement other functions without hardware replacements.

Detector Name

RAID-M Handheld Chemical Agent Detector

Manufacturer/Distributor and Contact Information

Bruker Daltonics Inc.
40 Manning Park
Billerica, MA 01821
Tel: 978-663-3600

Technique Description

This detector is a handheld ion mobility spectrometer. It can quickly detect very low concentrations due to highly sophisticated software.

Chemical Detection Capability and Performance

The detector identifies nerve, blister, and blood agents, and some TICs. Sensitivity is expected to meet/exceed current JSOR requirements.

Other Features

Audio and visual alarm indicators. Interface for connecting to computer networks. Additional compound detection can be programmed on request. Automatic purge mode to prevent oversaturation and achieve fast recovery time.

Detector Name

RAID-S® Stationary Rapid Alarm and Identification Device

Manufacturer/Distributor and Contact Information

Bruker Daltonics Inc.
40 Manning Park
Billerica, MA 01821
Tel: 978-663-3600

Technique Description

This device is a stationary ion mobility spectrometer for detecting CWAs and TICs through mode changes. Automatic polarity switching permits detection of all CWAs continuously. Sophisticated software prevents false alarms due to interfering substances. Specially designed for continuous operation for up to 1 year before maintenance.

Chemical Detection Capability and Performance

The detector identifies nerve, blister, and blood agents and some TICs. Sensitivity is expected to meet/exceed current JSOR requirements.

Other Features

Audio and visual alarm indicators. Interface for connecting to computer network. Additional compound detection can be programmed on request. Automatic purge mode to prevent oversaturation and achieve fast recovery time.

Detector Name

Sabre2000

Manufacturer/Distributor and Contact Information

Smiths Detection
2202 Lakeside Blvd.
Edgewood, MD 21040
Tel: 410-510-9100

Technique Description

This instrument is a handheld CWA, TIC, explosive, and narcotics detector that uses IMS. It has selectable operating modes including vapor or particulate sample detection.

Chemical Detection Capability and Performance

The device can detect nerve, blister, blood, and choking agents at JSOR/IDLH concentration levels with near real-time detection. It can also detect many TICs, explosives, and narcotics. It has a surface wiper that can collect persistent compounds for subsequent introduction to the detector for analysis in its particulate detection mode.

Other Features

Audio and visual alarm indicators. Can be upgraded to detect new compounds. Consumables include occasional desiccant cartridge changes. Suitable to serve as a monitor in CWA or Homeland Defense scenarios.

Domestic Preparedness Program Report

Domestic Preparedness Program: Testing of Sabre2000 Handheld Trace and Vapor Detector Against Chemical Warfare Agents Summary Report, August 2001 (www2.sbccom.army.mil/hld/ip/sabre_2000_download.htm).

CHAPTER 7

Flame Photometry

Flame photometry is an atomic spectroscopy technique based on the light emission phenomenon of excited atoms or clusters as they return to lower energy states. It is a very important CWA detection technique that has been successfully used for a long time. The detection of the characteristic light emission by excited sulfur and/or phosphorus atoms or clusters is used to identify CWAs. Most classic CWAs, such as the G-agents, V-agents, and sulfur mustard (HD), contain phosphorus and/or sulfur in their molecules as discussed in Chapter 2. Detectors based on atomic spectroscopy technology can also be used in detecting many TICs that contain phosphorus, sulfur, arsenic, or other atoms that produce characteristic light emissions.

Flame photometric detectors (FPDs) have been in use for as long as gas chromatographic and liquid chromatographic detectors. They are arguably the most useful analytical detectors for CWA studies. Since many of the chemicals of greatest concern, such as pesticides, automobile exhaust, and petroleum products, contain sulfur or phosphorous, FPDs are widely used in research and manufacturing operations. The CWA limit of detection without the addition of a gas chromatographic column is in parts per billion to parts per million.

An FPD is commonly connected to a gas chromatographic instrument to permit separation of chemicals in the sample for qualitative and quantitative analyses by benchtop instruments. The use of a gas chromatography (GC) column provides a means for increased selectivity and possible identification through respective elution times for diverse target compounds. Standard methods using gas chromatography coupled with a flame photometric detector (GC-FPD) were developed for and adopted by the Environment Protection Agency (EPA) and other government agencies for environmental studies. Needless to say, the GC-FPD method has been one of the most useful in determining CWA concentrations.

FPD instruments for field operations have portability and fast real-time analysis. Analyzing a vapor sample with a handheld FPD does not require sample preparation, as it draws a sample directly from the surrounding air and provides a content analysis instantly. For example, the AP2C® instrument made by Proengin is designed for direct air sampling. It can simultaneously and quickly detect both phosphorus- and sulfur-containing compounds, as the sample does not pass through a GC column. If the air contains phosphorus or sulfur, the detector responds within seconds. The AP2C is the only handheld FPD available at present.

FPD for CWA detection is considered semi-selective in the sense that it responds to all sulfur/phosphorus-containing chemicals. Without a means of compound separation, the FPD detector is not capable of distinguishing whether the detected signal derives from a CWA or another substance containing phosphorous and/or sulfur.

7.1 PRINCIPLE OF OPERATION

You may have observed bright yellow flames in the barbecue grill when salt or marinade drops onto the charcoals. The yellow flame emission is characteristic of sodium atoms. In contrast, when potassium salt is burned, the characteristic color emission is reddish. Each element emits a particular color when burned in an open flame. FPD is based on this principle.

An atom can be simply described as a nucleus with electrons orbiting around it. Roughly speaking, electrons revolve around the nucleus at different orbits, which reflect different atomic energy levels. The energy of an electron in a given orbit is a constant. Thus, the atom can be said to possess a certain energy level based on a certain orbit.

For example, hydrogen is the first element in the periodic table and each atom has only one electron. Figure 7.1a shows a few possible energy levels for a hydrogen atom. When the electron is in its lowest-energy orbit, the atom is said to be at "ground state" (Figure 7.1b), which is also the most stable. The atom will stay at this ground state as long as it is not stimulated by external energy, such as absorbing energy from a passing photon that will propel it to a higher energy, or excited, state (Figure 7.1c). At this excited state, the atom is unstable and has th e tendency to return to its lower ground-state energy level whenever possible. As it returns to lower energy levels, the absorbed energy is released by emitting a photon with a certain amount of energy (Figure 7.1d). The energy associated with this release of photon is equal to the difference between an atom's energy at an excited energy level and the ground-state energy level, which is the same as the energy absorbed initially.

The energy of the released photon (E) from the excited hydrogen atom from energy level 2 (n = 2) to energy level 1 (n = 1) (Figure 7.1d) is therefore

$$E = 10.19 - 0 = 10.19 \text{ (eV)} \tag{7.1}$$

The wavelength of the photon released from an atom depends on the energy of the photon (E) and can be calculated with the following equation:

$$\lambda = \frac{hc}{E} \tag{7.2}$$

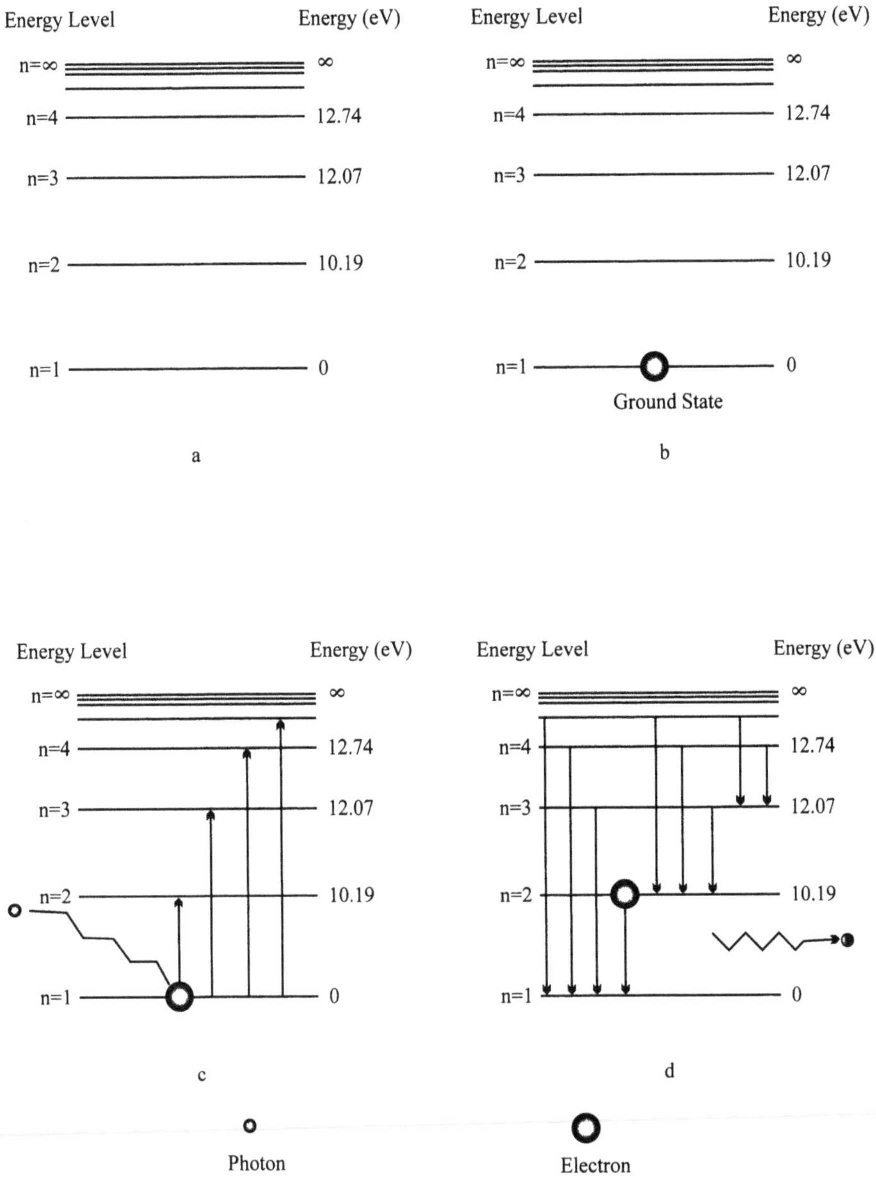

Figure 7.1 Energy levels of hydrogen atom (a); ground state (b); excited state via photon (c); and release of photon by excited electron (d).

where λ is the wavelength in nanometers (nm, 10^{-9} m); h is Planck's constant (h = 4.136×10^{-15} eV·sec); and c is the velocity of light (3×10^{17} nm/sec). From Equation 7.2, the relationship between the frequency (ν) of a photon and its energy can be calculated as

$$E = \frac{hc}{\lambda} = h\frac{c}{\lambda} = h\nu \tag{7.3}$$

Table 7.1 Frequencies and Wavelengths of Electromagnetic Waves

Common Name of Wave	Gamma Rays	X-rays	Ultraviolet	Visible	Infrared	Microwave	Radio
Wavelength	← Shorter						Longer →
(nm)	<0.01	0.01–1	1–400	400–700	700–10^5	10^5–10^8	>10^8
Frequency	← Increase						Decrease →
(Hz)	>3 × 10^{19}	3 × 10^{19} – 3 × 10^{17}	3 × 10^{17}– 7.5 × 10^{14}	7.5 × 10^{14}– 4.3 × 10^{14}	4.3 × 10^{14}– 3 × 10^{12}	3 × 10^{12} – 3 × 10^9	<3 × 10^9
Energy of	← Higher						Lower →
proton (eV)	>10^5	10^3–10^5	3–10^3	2–3	0.01–2	10^{-5}–0.01	<10^{-5}

When the hydrogen atom returns to energy level 1 (ground state) from energy level 2 (Figure 7.1d), it emits a photon of the following wavelength:

$$\lambda = \frac{hc}{E} = \frac{4.136\times10^{-15}(eV\cdot s)\times3\times10^{17}(nm/s)}{10.19(eV)} = 122(nm) \qquad (7.4)$$

Light at a wavelength of 122 nm falls in the UV range of the light spectrum (Table 7.1).

The number of electrons that can be excited depends on the element. An element's atomic number (listed in the periodic table) is determined by the number of protons in its nucleus, and the number of protons is equal to the number of electrons in the atom. For example, a phosphorous atom has 15 electrons, and a sulfur atom has 16 electrons. The electrons revolve around the nucleus at different orbits according to the energy each possesses when the atom is at ground state. Because more than one electron can be excited, and each of the electrons can be excited to different energy levels, when these excited electrons return to respective lower energy levels, the light they emit will be of different wavelengths, which in turn generates a light spectrum of different wavelengths. For the hydrogen atom, the light wavelengths released when its energy level falls to ground level are within the UV range, and visible light is released when its energy drops from a higher energy level to the n = 2 level.

As discussed earlier, most of the CWAs contain phosphorus or sulfur. When these molecules are broken down in a hydrogen–air flame, the released phosphorus and sulfur are excited. The excited sulfur (in the form of S_2, and designated as S_2^*) and phosphorous (in the form of HPO, and designated as HPO*) release light in respective characteristic wavelengths when they return to lower energy levels. FPDs identify the released characteristic wavelength of sulfur and/or phosphorous and thus determine whether a sample contains sulfur and/or phosphorous components.

Figure 7.2 is a block diagram showing the operational principle of FPDs. After the sample is introduced, the flame decomposes the molecules. S_2 and HPO from molecule decomposition absorb energy and become excited in the hydrogen–air flame. The excited S_2^* and HPO* then emit characteristic light when they return to their ground states:

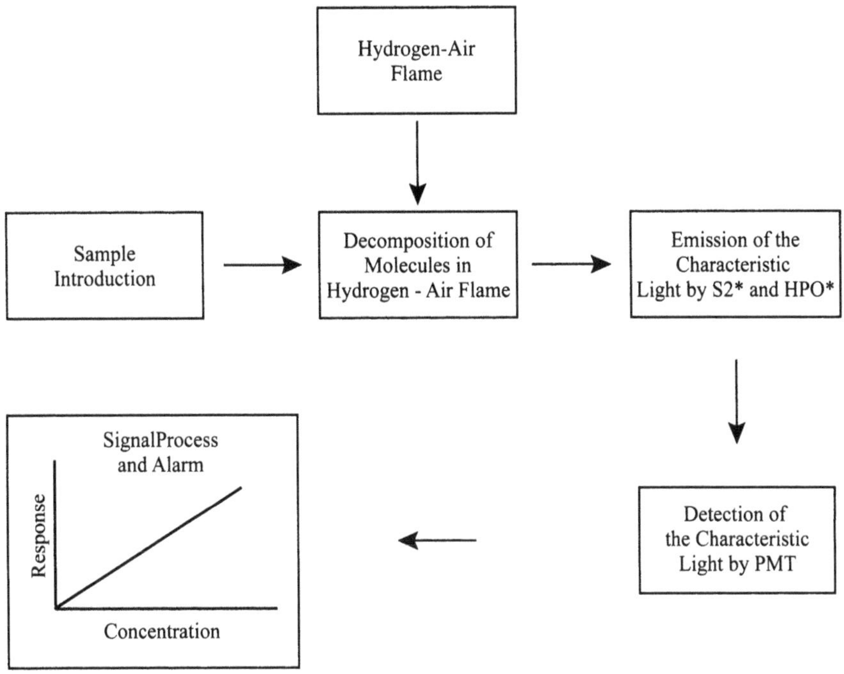

Figure 7.2 Block diagram of FPD analysis procedure.

$$S_2^* \rightarrow S_2 + h\nu \tag{7.5}$$

$$HPO^* \rightarrow HPO + h\nu \tag{7.6}$$

In Equations 7.5 and 7.6, hν represents the energy of the released photon. To achieve specificity, the FPD identifies the main emission wavelength of S_2^* (394 nm) and HPO^* (526 nm) with the use of filters to block photons of other wavelengths released in the burning process. The photons that pass through the light filter then strike the photomultiplier tube (PMT). The PMT monitors light intensity to produce the detection signal. The intensity of the light released is proportional to the number of S_2^* or HPO^* atoms, which is in turn related to the concentration of sulfur or phosphorous species in the sample. Thus, the greater the target chemical concentration, the greater the detection signal intensity. With the use of standard calibration curves, concentration of targeted chemicals in the sample can be correlated.

7.2 INSTRUMENTATION

Figure 7.3 shows a schematic diagram of a typical FPD. The detector consists of a hydrogen inlet, air inlet, sample inlet, burner, thermal filters, light filter, PMT, signal processor, and exhaust chimney. A source of hydrogen is required to produce the hydrogen–air flame for sample ionization.

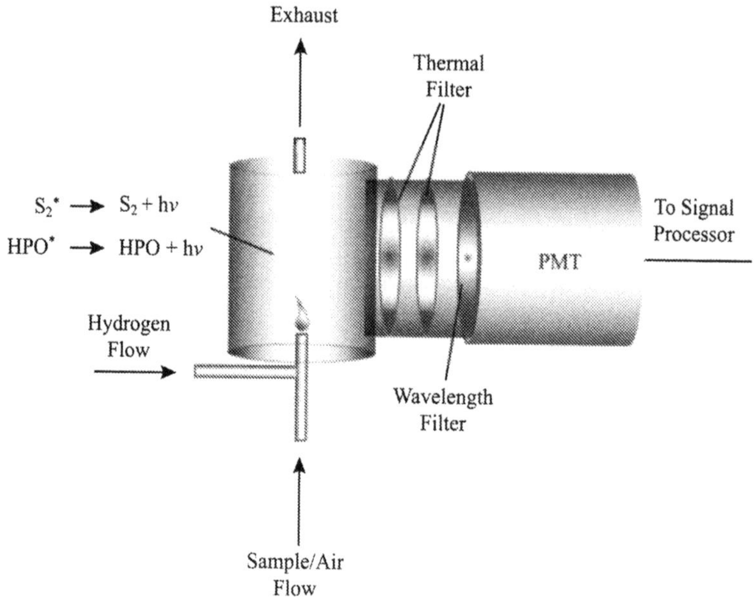

Figure 7.3 Schematic diagram of FPD.

7.2.1 Direct Sample Introduction and GC Interface

The direct sample type of FPD draws the sample via a pump into its combustion chamber at a controlled rate. The sample mixes with the hydrogen and burns at the tip of the burner. CWAs or TICs that contain sulfur and/or phosphorus are decomposed and generate signals. The detection occurs quickly, and is considered as real-time detection. However, without additional identification methods, the FPD cannot distinguish among the signals generated by multiple chemicals to identify the substances (i.e., CWAs versus other chemicals). The burning of other chemicals such as pesticide compounds can produce similar sulfur and/or phosphorus signals. (FPD has been used for detecting pesticide levels for a long time.) Sulfurous and phosphorous chemicals may also be found in smoke emissions during and after fires. Fortunately, gas chromatographic elution is a very powerful tool that can be combined with the FPD to enable compound identification.

Basically, gas chromatography separates chemicals in vapor based on each chemical's distinct distribution coefficient between the stationary and mobile phases. A column containing the stationary phase is the heart of the technique, and the carrier gas (mobile phase) is used to carry the chemicals through the column. (A complete discussion of gas chromatography is beyond the scope of this book. Numerous excellent publications on gas chromatography are available, and we have included three references in the bibliography at the end of the book.)

For the FPD, a GC column can be used to separate chemicals in the sample before they enter the mixing and combustion chamber according to their respective elution time. The combination of retention times determined by GC and the presence of sulfur and/or phosphorous signals enables the possible identification of targeted chemicals. Identification can only be considered as possible because many substances could have similar retention times through the GC column, which in turn may yield detection peaks by the FPD. However, such combinations occur infrequently. Therefore, the frequency of false alarms is drastically reduced. The drawback of the addition of a GC in front of the FPD is that analysis time increases. Consequently, detection cannot be considered as "real time."

7.2.2 Flame and Hydrogen Source

7.2.2.1 Hydrogen–Air Flame

A hydrogen–air mixture is ignited to produce the flame. The hydrogen–air flame, at temperatures of 2000°C to 2100°C, breaks down the incoming organic chemicals and provides the energy for exciting the electrons. This temperature is sufficient to excite S_2 and HPO without ionizing them. The hydrogen–air flame is also used in the flame ionization detection technique, which is discussed in Chapter 11.

7.2.2.2 Hydrogen Source

While benchtop instruments can use pressurized tanks as the hydrogen source, carrying pressurized hydrogen is very inconvenient in field applications, and many believe tanks in the field present a major safety hazard. Alternatively, small hydrogen alloy storage devices that contain reasonable amounts of hydrogen (in the form of a metal hydride) are available, and these can be used safely in the field.

Alloys such as Mg, Mg_2, Ni, FeTi, and $LaNi_5$ are capable of absorbing and releasing considerable amounts of hydrogen gas in and from relatively small quantities of hydride powder. These alloys can absorb and hold large amounts of hydrogen by bonding with hydrogen to form hydrides (MgH_2, Mg_2NiH_4, $FeTiH_2$, and $LaNi_5H_6$) under 1 atm. The bonded hydrogen can be released as needed. The advantage of these types of storage devices is their relatively large capacity for hydrogen absorption in small containers. The alloys can hold more hydrogen atoms than liquid hydrogen of the same volume. For instance, it has been reported that a cubic centimeter of LaNi alloy can absorb 5.5×10^{22} hydrogen atoms. The volume of a mole of hydrogen gas (6.02×10^{23} hydrogen molecules, or $2 \times 6.02 \times 10^{23}$ hydrogen atoms since one hydrogen molecule consists of two atoms) is 24.45 L under 1 atmospheric pressure at 25°C. Therefore, the volume of 5.5×10^{22} hydrogen atoms is approximately equivalent to 1 L of hydrogen gas at room temperature. For comparison purposes, 1 cm^3 of liquid hydrogen contains 4.2×10^{22} atoms, and 1 cm^3 of hydrogen gas contains only 0.99×10^{22} hydrogen atoms at barometric pressure 200. Another feature of the hydrogen alloys is the ability to absorb and release hydrogen gas for many cycles without deteriorating. The release rate of the absorbed hydrogen can be controlled by temperature or pressure regulation. The use of absorbed hydrogen has made hydrogen handling much more convenient and less hazardous than using cylinders of compressed hydrogen. The current AP2C detector uses a small cartridge of alloy-absorbed hydrogen, which contains enough hydrogen for efficient field applications.

The hydrogen enters the combustion chamber and mixes with the air sample from the air inlet to form the hydrogen–air flame. Sample air may come from the GC column if components are separated before the sample enters the detector. The flame decomposes chemicals contained in the sample. In the case of CWAs in a sample, the released and excited sulfur and/or phosphorus emit photons of a characteristic respective light spectrum for detection through the use of special light filters.

7.2.3 Signal Detection

The excited sulfur or phosphorus atoms emit characteristic light when they are returning to the stable ground state. This emitted light passes through two quartz windows and optical filters before reaching the PMT. Electrical signals are generated by the PMT and interpreted by a microprocessor to determine sulfur or phosphorus concentration in the sample.

7.2.3.1 Thermal and Wavelength Filters

The burning of the hydrogen in the air generates a large amount of heat. If the heat is permitted to reach the PMT, undesirable background noise can render the system useless. Heat will also shorten the useful life of the photometer. Heat from a flame is usually released in the form of infrared light. To reduce the effects of heat reaching the PMT, quartz plates are used as thermal filters to block a large portion of infrared light while allowing other light spectra such as UV and visible light to pass through. The quartz filters are located in front of the optical wavelength filters. The use of two quartz filters provides more effective heat shielding.

The light released by S_2^*, HPO^*, and other substances passes through the quartz thermal filters and then reaches the wavelength filters; only light of specific wavelengths can pass through these filters. For example, a 394-nm filter is used for sulfur, and a 526-nm filter is used for phosphorous detection. A filter change is generally required to put the instrument in the correct mode for detection of a targeted chemical. More sophisticated instruments may contain a dual FPD, where one of the FPD uses the 394-nm filter for sulfur detection and the other uses the 526-nm filter for phosphorus detection. The dual FPD permits simultaneous detection of both elements, which in turn enables detection of nerve and blister agents simultaneously in CWA detection applications. However, changing filters or employing dual detectors is impractical for field applications.

Field instruments like the AP2C incorporate both sulfur and phosphorus filters within the same detector. The user switches back and forth between the two types of filters to permit measuring characteristic light emissions from both elements.

7.2.3.2 PMT

After the photons emitted by sulfur or phosphorus pass through the wavelength filter, they are received by the PMT. The PMT is a vacuum tube that consists of a photocathode, several dynodes, and an anode to convert the incoming light into an

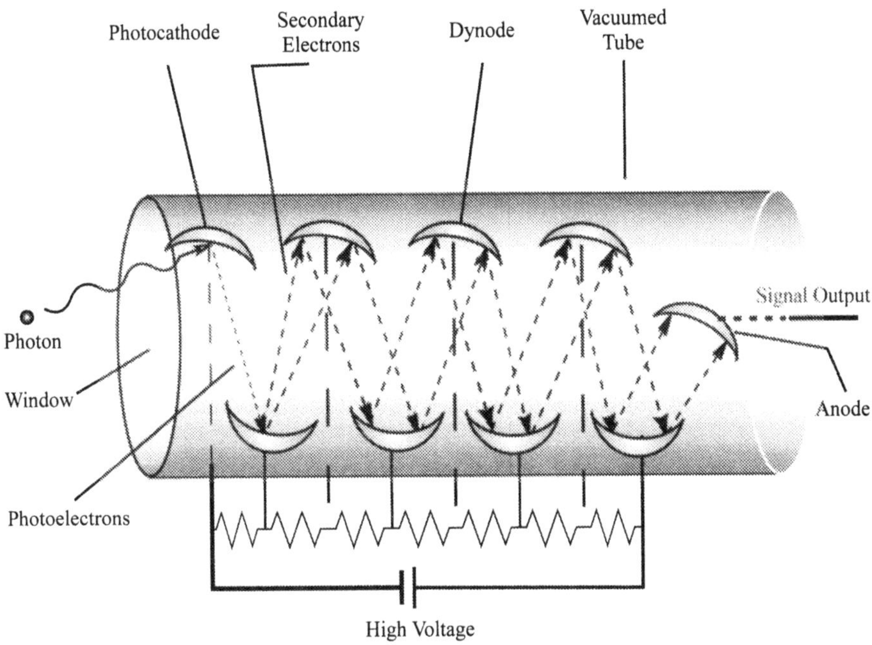

Figure 7.4 Schematic diagram of photomultiplier tube.

electrical signal (Figure 7.4). The light strikes the photocathode, which releases electrons called photoelectrons. These electrons are accelerated through an electric field to the first dynode. They, in turn, strike the second dynode and produce the release of many more electrons, which are referred to as secondary electrons. Each of these secondary electrons is then further accelerated to the next dynode and generates still more electrons. The process is repeated through each dynode. The number of electrons generated by a single electron through the photocathode is thus increased through several dynodes and multiplied in the tube. The number of electrons received by the anode for each photoelectron could be increased by 10^4 to 10^5 depending on the photomultiplier structure. The anode receives these electrons and generates the electric signals for the microprocessor to process.

7.2.3.3 Chemical Identification

Since only the characteristic light wavelengths generated by HP* or S_2^* are permitted to enter the PMT, the detection signal is considered to be an indication that the sample contains sulfur and/or phosphorous compounds. Such direct sampling results do not permit precise substance identification. As discussed previously, detection indicates only that the sample consists of substances that contain sulfur and/or phosphorus, which may or may not be targeted toxic chemicals. The AP2C is an example of this principle of detection.

When FPD is coupled with a GC column, the detected signals generated by the eluting chemicals can provide a means for compound identification. The retention time of an agent in the GC column is a function of temperature and carrier gas pressure together with the properties of the chemical and the column. The retention time for each agent or chemical of interest is obtained by running standard samples at a defined temperature and pressure with a certain type of column. Therefore, the identification of an agent via a GC-FPD is based on two factors: retention time and presence of sulfur and/or phosphorus. This drastically reduces the false positive alarm rate. The Miniature Continuous Air Monitoring System (MINICAMS®) is an example of using the GC mode successfully together with FPD for CWA detection. In addition to the GC mode, the MINICAMS also uses sample concentration and thermal desorption to increase its sensitivity for vapor detection.

The wastes produced through the FPD processes, which include water generated by burning hydrogen in air and decomposed substances from the sample, are removed from the detector via a chimney exhaust. The toxic sample drawn into the detector is decomposed, and therefore detoxified, when it is burned in the hydrogen flame.

7.3 TECHNICAL SPECIFICATIONS

7.3.1 Selectivity

Currently, FPDs are made to detect sulfur and phosphorous compounds through specific wavelength filters, and are able to detect respective CWA compounds such as the G-agents, mustard, and certain TICs, including PH_3, SO_2, and H_2S. Since other chemicals that do not contain these two elements do not emit these characteristic wavelengths, they do not interfere with detection. With the recent interest in detection of TICs, FPDs are of potential usefulness because various light filters can be installed to permit detection of other elements as well.

As discussed above, when the detector reports sulfur or phosphorus, there is no way to know if the detected substances are CWAs, TICs, or other nontargeted chemicals. Specificity can only be improved through use of a GC column, which will increase sample analysis time but permit possible compound identification and reduce false positive alarms.

7.3.2 Sensitivity and LOD

FPD is very sensitive for sulfur and phosphorous compound detection. The reported limit of CWA detection is at the parts-per-billion level, which is lower than the IDLH levels set for all CWAs and TICs.

7.3.3 Response Dynamic Range

The concentration–response relationship, unlike ion mobility spectroscopy detectors, is usually linear. For phosphorus, this relationship remains linear over a large concentration range. However, since sulfur exists in the flame as S_2, the concentration–response relationship is a second-order one. Response linearity is related to the square root of sulfur concentration. The overall dynamic range for both sulfur and phosphorous are over 10^4 or even 10^5.

7.3.4 Quantitative Analysis Capability

FPDs such as MINICAM use the intensity of signals generated by sulfur or phosphorous to determine target chemical concentration in the sample. They are widely used in laboratories for determining concentrations of CWAs in samples and have demonstrated excellent quantitative analysis capability. For field applications, responses are presented in a semiquantitative manner for simplicity. As an example, the AP2C detector uses five light bars to indicate detected concentration levels.

For quantitative analysis, it is important that background signals are corrected to reduce interference from background noise. This may be achieved by using two reference optical filters to monitor the light with wavelengths in the neighborhood of the light emitted by sulfur or phosphorus. These filters reveal the background intensity near the sulfur or phosphorous wavelengths that could add signals to the detector. The signals are then compensated for by the microprocessor to minimize the noise effect.

7.3.5 False Alarm Rate

As discussed above, FPD is made specifically selective to sulfur and phosphorus components for CWA detection purposes. Thus, other targeted and nontargeted sulfur and phosphorous components may trigger false positive alarms. False positive alarms may also be caused by interfering chemicals that emit photons at the same or similar wavelengths as those of sulfur or phosphorus.

7.3.6 Response Time

Detection of CWAs with FPD can be considered real time by instruments designed for direct air analysis. Air is continually being drawn through the detector and analyzed in real time. Detection time will increase if a GC column is used to separate sample chemicals, and is determined by the elution time of the targeted chemical through the column. GC column use will generally require sample preconcentration to produce a sample size adequate for detection.

7.3.7 Other Specifications

FPDs are usually designed as very rugged devices, and require very little maintenance. The PMT has quite a long life, and can be changed easily when replacement is necessary. In general, because the hydrogen–air flame used in this type of detector produces very high temperatures and burns clean, very little residue in the combustion chamber is expected. However, if the detector has been used to analyze samples with high organic matter content for a long period of time, it is possible that carbon may coat the thermo filter and sensitivity will be reduced. Maintaining the proper hydrogen–air mixture is the key to produce consistent flame temperature, and thus long-term consistent analytical sensitivity.

7.4 PULSED FLAME PHOTOMETRY

A relatively new technique called pulsed flame photometry has been developed and commercialized for benchtop instruments. It is based on the same fundamental principles as the FPD, except that the flame is pulsed through the use of various flow patterns and flow rate control. Hydrogen combines with air sample flows into a small, pulsed-flame chamber that contains a continuously heated wire igniter. Here the combustible mixture is ignited and propagated back to the gas source and self-terminated within a few milliseconds. The sample in the ignition chamber continues to produce additional ignition for a few hundred milliseconds. This ignition-propagation-termination cycle continues in periodic pulsed fashion.

In a pulsed flame photometric detector (PFPD), the combustion of hydrocarbon molecules is fast and irreversible, and heteroatom species such as S_2, HPO, and HNO emit light after the flame is extinguished and thus under cooler temperatures. Consequently, their respective emissions can be electronically gated and separated from the hydrocarbon emission. Thus, PFPD can provide selectivity against hydrocarbon interference during detection analysis. PFPD sensitivity was reported to be superior to FPD. Moreover, N and As could be also detected. The PFPD is currently available for use in benchtop instruments, such as the MINICAMS from O. I. Analytical and other GC detector manufacturers.

The PFPD has much lower LOD and better sensitivity than FPD technology. The MDL of sulfur and phosphorous is in the 10^{-13} to 10^{-14} g/sec area (see Aviv Amirav et al. on www.tau.ac.il/chemistry/amirav/pfpd.shtml). In addition, the PFPD can simultaneously detect several element combinations using only one photomultiplier. Element identification can be achieved through the dual-gate-response ratio method. This detection technique is capable of identifying sulfur, phosphorous, nitrogen, arsenic, and many other elements with no interference from hydrocarbons. Thus, the PFPD would be very suitable for TIC detection.

The author has proposed to develop and commercialize the handheld Fast GC-PFPD model combines sample concentration, thermal desorption, GC separation together with highly selective PFPD detection. This device promises high sensitivity, selectivity, and specific CWA and TIC detection. Unfortunately, as of this writing, it was not commercially available. The Fast GC-PFPD operates in a fashion similar to the benchtop MINICAMS equipped with the PFPD detector. It differs from the MINICAMS in that it does not require various gas cylinders, but rather a metal hydride cartridge. Analysis time would be 35 sec for a full-cycle analysis. It can also be operated under the "sniff" mode. The LOD for the Fast GC-PFPD is expected to be 0.003 $\mu g/m^3$ organophosphorus and 0.2 $\mu g/m^3$ organosulfur.

7.5 APPLICATIONS

The MINICAMS-FPD (Figure 7.5) is a prime example of how the technology has impacted on studies of CWAs. MINICAMS is the analytical instrument of choice in CWA laboratories. Given that the FPDs detect both sulfur and phosphorus effectively, the MINICAMS-FPD covers nearly all commonly tested chemical agents.

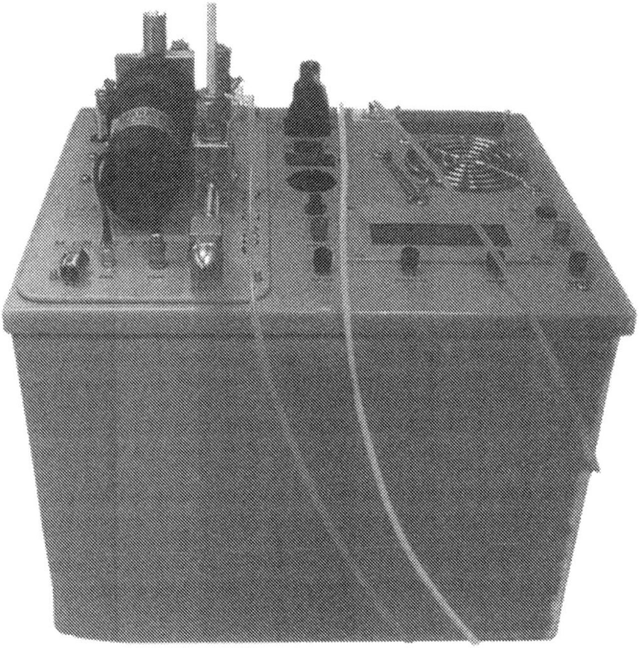

Figure 7.5 MINICAMS Detector. Photo courtesy of K. Y. Ong.

MINICAMS-FPD has also been placed at various sites where toxic chemical exposure may occur to serve not only as an analytical tool but also as a continuous safety monitoring device. The usefulness of this device has been repeatedly demonstrated in monitoring the potential escape of CWA vapor from engineering controls in a workplace. It can detect CWA vapors at sub-AEL concentrations.

GC manufacturers such as Tracor and PerkinElmer have developed FPDs. The detector is located behind a GC column to identify elution peaks after separation take place as the sample migrates through the column. Proper installation of optical filters permits the FPD to detect various target compounds. For CWAs, filters for phosphorus or sulfur detection are used for nerve agents and HD detection, respectively. Some users deploy two separate detectors, one installed with the phosphorus detection filter while the other uses the sulfur detection filter to achieve simultaneous detection of nerve agents and HD should the sample contain either or both.

Tracor has developed a CWA detector, the HyFED®, which draws an air sample directly into the FPD for analysis. The instrument requires connection to compressed hydrogen and air cylinders to provide the hydrogen flame for sample ionization. The instrument provides the means to sample continuously and analyze an airstream in real-time fashion. This detector has proven useful in monitoring vapor concentrations generated in laboratories. Because the targeted sample is known in laboratory conditions, there is no need to use a GC column for sample separation. Therefore, detection is direct and output instantaneous. Relative response signals can be correlated to provide a quantitative determination.

The French company Proengin Inc. (78210 Saint Cyr L'Ecole) developed the AP2C detector, a handheld device capable of simultaneously detecting nerve and blister

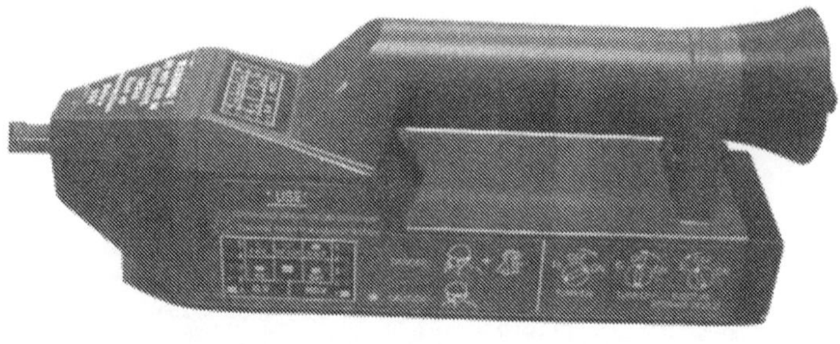

Figure 7.6 AP2C Detector. Photo courtesy of K. Y. Ong.

agents (Figure 7.6). According to the manufacturer's operating instructions manual, the operational temperature performance ranges from –32°C to 55°C. The instrument weighs approximately 2 kg, including the 7.3-V battery pack and hydrogen storage cartridge that provides hydrogen for approximately 12 hours of operation when fully charged. Start-up time is less than 2 min at temperatures above 0°C, and up to 15 minutes at temperatures below 0°C. Heat is applied to the hydrogen cartridge to maintain effective hydrogen production at cold temperatures. The UC-AP2C® model includes the vapor monitor (AP2C) and the liquid or solid substance sampler (S4PE®). During normal vapor operation, sample air is drawn into the AP2C unit through the changeable "vapor-sampling" nozzle at a sampling rate of approximately 80 l/hr (1.33 l/min). The heat from the flame decontaminates the internal surfaces, which allows fast recovery times.

The instrument is sensitive and can quickly detect CW agents. It utilizes a turning wheel of light filters to alternatively sense lights emitted from phosphorus and sulfur as a sample is burned. The internal light filters are alternated automatically without the need of manually changing the detection mode. Therefore, the device is capable of detecting nerve and blister agents simultaneously. When target substances are detected, it indicates the detection of phosphorus (i.e., nerve agent) or sulfur (HD). Because VX molecules consist of both sulfur and phosphorus atoms, its characteristic flame emission includes both the sulfurous and the phosphorous characteristics. When VX is detected, both the phosphorus and sulfur light indications on the AP2C panel will show. Because of lesser sensitivity toward sulfur detection than phosphorus detection, lower concentrations of VX may trigger only the phosphorus detection response.

The detected sample concentration is reported through the use of five rows of LED bars corresponding to relative concentration of HPO_2 and S_2. More bars lit indicate a higher detected concentration. The operator's manual states that the hazard threshold for unprotected humans is reached as soon as the first red indicator light flashes on.

The AP2C also has the capability to detect liquid surface contamination using the S4PE accessory. The AP2C has been tested under the Domestic Preparedness program. Interested readers can access the full report on the Homeland Defense website.

7.6 CONCLUSION

In summary, the flame photometric technique is based on the wavelength (or frequency) of light emissions produced as a substance is burned when it passes through a hydrogen–air flame. This technique uses the hydrogen–air flame as its energy source to decompose the sample and excite sulfur and phosphorous atoms. The energy loss produces photons that emit light when the excited sulfur or phosphorous atoms return to their ground state. The PMT multiplies the photoelectrons and produces signals. Since sulfur and/or phosphorous is commonly found in CWAs and some TICs, FPD is therefore capable of detecting CWAs and some TICs. It is also capable of detecting several other elements using various light filters. The microprocessor converts the signal that can be related to sample concentration. Both handheld and benchtop instruments are available.

The desirable features of the flame photometer are sensitivity, stability, and relatively large linear dynamic range. FPDs also require very little maintenance, and are ready to use in a very short time. An FPD's response to phosphorous is a first-order relationship, while its response to sulfur components is second order because sulfur is detected as S_2. Analysis by an FPD without a GC column is considered real-time detection. Unfortunately, without GC involvement, the detector cannot provide possible compound identification due to its generic nature. The addition of GC capability permits identification of sample components that contain targeted elements, and reduces the false alarm rate.

7.7 FACT SHEETS ON SELECTED PHOTOMETRY-BASED DETECTORS

Detector Name

AP2C Vapor Detector

Manufacturer/Distributor and Contact Information

Proengin Inc.
405 N.E. 8th Street
Fort Lauderdale, FL 33304
Tel: 954-760-9990

Technique Description

A handheld device using flame photometric spectroscopy, this detector is operable under field environments. It is designed for detection of selected TICs and nerve and blister agents at or below JSOR or IDLH levels through the use of chopping light filters to modulate light emission signals.

Chemical Detection Capability and Performance

This device does not have inlet filter. Real-time detection response occurs in seconds. Start-up time is less than 2 min except in very cold temperature conditions. Use of a metalhydride hydrogen cartridge is required for operation.

Other Features

LED bars are used for visual alarm indicators. Audio and remote alarms are also available, as is interface for network connection. Multicompound detection capability is based on elemental content of the substance (P or S). A surface probe (SP4) is available to sample surface contamination by vaporizing the collected substance through heating for detection by the AP2C unit, which permits detection of persistent compounds. Hydrogen flame breaks up toxic compounds into harmless elements.

Domestic Preparedness Program Report

Domestic Preparedness Program: Testing of Detectors Against Chemical Warfare Agents — Summary Report UC P2C Portable Chemical Contamination Control Monitor Collective Unit, May 2001 (http://hld.sbccom.army.mil/ip/reports.htm#detectors)

Detector Name

Chemical Weapons Agent Detection & Confirmation System

Manufacturer/Distributor and Contact Information

Agilent Technologies
2850 Centerville Road
Wilmington, DE 19808
Tel: 800-227-9770

Technique Description

A semiportable or stationary thermal desorption/gas chromatographic type of instrument, this device is operable in field environments in a mobile laboratory unit. It is designed for identification of a long list of TICs and nerve and blister agents down to or below AEL levels through the use of different modules equipped with various signal detectors.

Chemical Detection Capability and Performance

The signal detectors that identify the various compounds are FPD and MSD. Because it is a sample-concentrating device using thermal desorption and GC separation/detection, process times are in minutes, according to the desired detection level, rather than in seconds. Therefore, it is not suitable for gross-level detection.

Other Features

Interface for network connections. Additional compound detection can be programmed. Device can be networked. Multicompound detection capability according to the specific signal detector used. This device is a proven technology capable of monitoring to sub-AEL concentrations in the air.

Detector Name

MINICAMS Model FM-3001 (Miniature Continuous Air Monitoring System)

Manufacturer/Distributor and Contact Information

CMS Field Products
O.I. Analytical
2148 Pelham Parkway, Bldg. 400
Pelham, AL 35124
Tel: 205-733-6900

Technique Description

A semiportable or stationary thermal desorption/gas chromatographic type of instrument with flame photometric detector, this device is operable in field environments. It is designed for the detection of a long list of TICs and nerve and blister agents down to or below AEL levels through the use of different modules equipped with various signal detectors.

Chemical Detection Capability and Performance

Requires multiple signal detectors to achieve detection of various compounds. Because it is a sample-concentrating device using thermal desorption and GC separation/detection, alarm times are in minutes, according to the desired detection level, rather than seconds. Therefore, it is not suitable for gross-level detection.

Other Features

Audio and visual alarm indicators. Interface for network connections. Additional compound detection can be programmed. Multicompound detection capability according to specific signal detector used. The MINICAMS is a proven technology capable of monitoring to sub-AEL chemical-agent concentration in the air.

Detector Name

TIMs Detector for Toxic Industrial Materials Detector

Manufacturer/Distributor and Contact Information

Proengin Inc.
405 N.E. 8th Street
Fort Lauderdale, FL 33304
Tel: 954-760-9990

Technique Description

Similar to the AP2C detector, this handheld device uses flame photometric spectroscopy detection. It is operable in field environments, designed for detection of TICs or TIMs at or below IDLH levels through the use of chopping light filters to modulate light emission signal detection. The unit is desensitized from the AP2C specially designed for TIC detection which does not require such high sensitivity.

Chemical Detection Capability and Performance

Sensitivity is expected to meet/exceed current IDLH requirements. Device does not have an inlet filter. Real-time detection response occurs in seconds. Start-up time is less than 2 min except in very cold temperatures. Use of hydrogen cartridge required for operation.

Other Features

Visual indicator of alarm through LED bars. Audio and remote alarms are also available. Interface for connecting to computer networks is available. Multicompound detection capability according to elemental content of substance. A surface probe (SP4) is available to sample surface contamination by vaporizing the collected substance through heating for detection by the AP2C unit, which permits detection of persistent compounds. Hydrogen flame breaks up toxic compounds into harmless elements.

CHAPTER 8

Infrared Spectroscopy

Gamma rays, x-rays, ultraviolet light, visible light, infrared light, microwaves, and radio waves are electromagnetic radiation waves. The only difference among these different types of radiation is wavelength or frequency range. In order, wavelength increases from gamma rays, x-rays, ultraviolet, visible, infrared, and microwaves to radio waves. Table 7.1 shows that the energy level declines as wavelength increases. Infrared (IR) wavelengths range from 0.78 μm to 1000 μm, where 1 μm equals to 10^{-6} m (commonly referred to as 1 micron). Infrared detectors use a chemical's IR radiation absorption characteristics to detect and identify it.

Sir Frederick William Herschel (1738–1822) discovered IR radiation in 1800. He found that temperature increased from the violet to the red part of the visible light spectrum. After noticing this pattern, he measured the temperature just beyond the red portion of the spectrum in a region where color was invisible. To his surprise, he found that this region had the highest temperature. Additional experiments showed that the higher temperature was from a form of light or radiation now known as infrared radiation. The IR wavelength is longer than that of the red end of the visible light spectrum and shorter than that of microwave radiation. Thus, IR photons have lower energy than visible light, but possess more energy than microwaves.

There are two major types of IR instruments: one type detects IR emitted from objects, such as IR cameras; and the other detects IR radiation absorbed by objects, such as IR spectroscopy. The uses of the IR technique in the analytical and detection field are based on IR absorption characteristics at specific wavelengths by the substances of interest.

Function groups in a molecule vibrate at certain frequencies that fall within the IR region. The molecule absorbs energy from IR waves that have the same frequencies, as the vibrating function groups are excited to a higher energy level. Specific function groups absorb IR radiation having specific frequencies. By determining the IR radiation frequencies being absorbed by the function groups composing the molecule, the molecular structure can be determined. The spectrum of a large molecule can be very complex, depending on the number of function groups in the molecule. Therefore, large IR spectra databases have been established for many chemicals. For example, the National Institute of Standards and Technology has a database comprised of 5228 infrared spectra. The identification of a substance and its structure can be determined through matching its IR absorption spectrum with spectra from the databases.

A simplified method is generally used for detecting certain groups of chemicals in a vapor sample. For field applications of CWA and TIC detection, the objective of course is to determine whether the sample contains the targeted chemical(s) rather than identifying the structure of these chemicals. The characteristic frequencies of targeted chemicals are known based on their function groups and confirmed in laboratory studies. Therefore, it may suffice to check the IR absorption using a few characteristic wavelengths to achieve the detection objective. This can be done by exposing the sample to IR radiation at several selected wavelengths that are characteristically absorbed by the targeted chemicals. This greatly simplifies the instrumentation design of IR detectors. We will focus on detection aspects rather than structure identification using various IR technologies.

Detectors based on IR technique have the advantage of reasonably high sensitivity and fast detection. More sophisticated instruments can offer better selectivity. IR is a nondestructive technique and requires minimal sample preparation. Remote detection of chemical clouds can also be made using the infrared emission technique. Environmental conditions, especially humidity, may significantly affect its performance, however.

8.1 PRINCIPLE OF OPERATION

Figure 8.1 shows a block diagram illustrating the principle of operation of IR detectors in general. A sample is introduced into the detector and exposed to IR irradiation. The irradiation can derive from a filtered IR source at a certain wavelength or an IR source that does not use filters to cover the entire IR wavelength range for different applications. Ideally, when the targeted chemical is present in the sample, IR absorption at characteristic wavelengths occurs. The intensity of IR absorption for those wavelengths changes according to the concentration of the targeted chemical. Signals are sent to the central processor for chemical identification and/or concentration determination.

8.1.1 Infrared Sources

IR wavelengths range from 0.78 μm and 1000 μm. Consequently, the frequency of an IR radiation is in the range of 3×10^{11} Hz to 3.8×10^{14} Hz. Table 7.1 shows the IR radiation region relative to other types of radiation in the electromagnetic spectrum. IR radiation can also be expressed in terms of wave numbers, or the number of full waves per centimeter. Wave numbers are expressed in number of waves per centimeter units (cm^{-1}). Unlike wavelengths, a wave number is directly related to its energy level. Higher wave numbers possess higher energy levels.

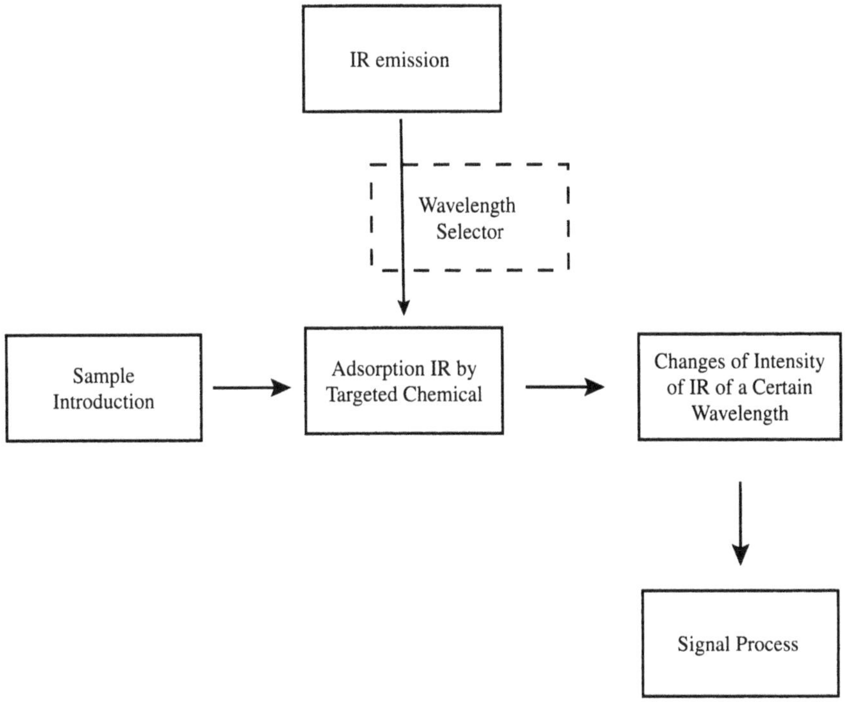

Figure 8.1 Block diagram of IR detector.

IR radiation is further subclassified into three types according to wavelength. The near visible IR (NVIR) wavelengths range from 0.78 μm to 2.5 μm, equivalent to wave numbers ranging from 12,800 to 4000 cm^{-1}. The wavelengths of middle IR (MWIR) range from 2.5 μm to 50 μm, equivalent to wave numbers 4000 to 200 cm^{-1}. Far IR (FIR) covers wavelengths ranging from 50 μm to 1000 μm, equivalent to wave numbers 200 to 10 cm^{-1}. IR with wavelengths between 2.5 μm and 15 μm (670 to 4000 cm^{-1}) are the most frequently used in detection applications.

In the real world, any object will always have a temperature above absolute zero (−273.15°C or 0°K). Because the vibration of atoms or function groups in a molecule emits electromagnetic waves having frequencies in the range of IR radiation, all objects produce infrared radiation. Night-vision cameras are examples of remotely detecting objects based on IR emissions. While passive, remote-sensing IR monitoring devices make use of objects' IR radiation for detection, active types of IR devices require an IR source of their own to irradiate the sample after which the absorbed irradiation is measured. The primary source of IR is made available through the use of heat or thermal radiation. An inert solid that is heated to a temperature of 1500 to 2000°K will continually emit the necessary IR for a detector. However, the intensity of the generated IR at different wave numbers varies. Maximum intensity occurs between 5000 cm^{-1} and 5900 cm^{-1}, with gradual decline toward smaller wave numbers and sharp decline toward higher wave numbers.

8.1.2 Molecular Absorption

Organic molecules are composed of various atoms and function groups that are connected to each other through different types of bonds. Each of these bonds vibrates in a specific manner at a certain frequency. Figure 8.2 illustrates some of these vibration modes, including stretching, in-plane bending, and out-of-plane bending. Stretching vibration is the rhythmic movement along an axis of a bond (Figure 8.2a). The bond length increases and decreases at a certain frequency characteristic of the specific bond. In-plane bending vibration is the rhythmic change in the bond angle or movement relative to the rest of the molecule within the plane of the molecule (Figure 8.2b). Similar to in-plane bending, out-of-plane bending vibration is referred to as the rhythmic change in the bond angle relative to the rest of the molecule, but the bending is out of plane with the rest of the molecule (Figure 8.2c). In various chemicals, the same characteristic function group vibrates in the same fashion and absorbs IR radiation of same frequency. For other substances, the same characteristic function group vibrates similarly and absorbs IR with the same or similar wavelength. For example, when the C-C bond vibrates in the stretching fashion, it absorbs IR of 2800 to 3000 cm^{-1}. This similarity is most apparent from the wave numbers greater than 1200 cm^{-1}. This range is called the group frequency region.

Different structures, however, may affect the vibration frequency of the characteristic function group, resulting in different absorption characteristics at the low wave-number range of less than 1200 cm^{-1}, known as the fingerprint region. This feature makes it possible to identify the function groups in a molecule based on the frequencies absorbed by the molecule when scanned through the entire IR spectrum.

When an object absorbs energy from IR radiation, the electromagnetic wave from the IR source is transformed into mechanical vibration of the molecules as an IR photon with same vibration frequency collides with the molecule. The intensity of the IR at that frequency declines after it passes through the sample. Measuring these changes during the scanning through the IR wavelength range produces the fingerprint of the substance.

Molecules can be grouped into symmetric and asymmetric molecules. It was found that the symmetrical molecules with symmetric vibration (e.g., N_2, O_2, CH_4, C_6H_6, and CCl_4) do not absorb IR radiation (however, symmetrical molecules with asymmetric vibration such as CO_2 will still absorb IR). This is the most fortunate and important feature that permits the use of IR techniques for atmospheric vapor–monitoring applications. Because the two major components of air, N_2 (78%) and O_2 (21%), are dipoles and symmetrical molecules, they do not absorb IR. Thus, their presence and abundance will not interfere with IR detection of other chemicals in the air.

8.1.3 Beer's Law

When an IR beam at a certain wavelength passes through a sample, the molecules that vibrate at the same frequency as the bombarding IR radiation absorb the radiation. This absorption decreases the intensity of the IR beam (Figure 8.3). The absorbance (A) value is a measurement of this loss in intensity, as follows:

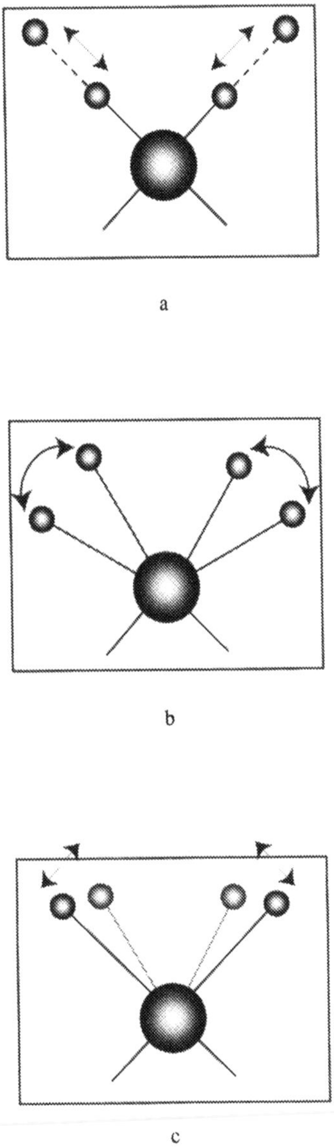

Figure 8.2 Examples of molecular vibration model.

$$A = \log_{10} \frac{I_0}{I} \tag{8.1}$$

where I_0 is the initial intensity of the incoming radiation at the wavelength emitted from the source to the sample, and I represents the radiation after passed through the sample. Beer's law states that the absorbance of radiation at a certain wavelength is linearly related to the total amount of molecules that absorb the radiation in the sample. The total number of molecules is directly related to the path length (L) and

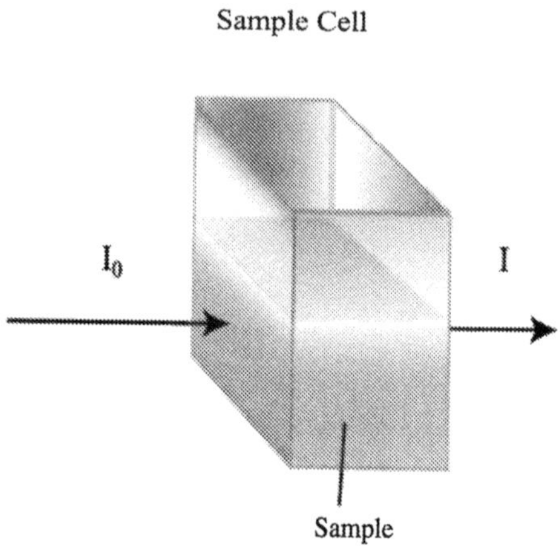

Figure 8.3 Beer's Law.

the concentration of the sample (C) that the IR beam travels through as expressed in Equation 8.2:

$$A = eCL \tag{8.2}$$

where e (in the unit of area/mass, such as square centimeter per milligram) is the molar absorptivity known as the extinction coefficient.

Most substances obey Beer's Law at low to moderate concentrations that still permit transmission of the IR beam. Higher percentages of IR transmitted through the sample produce lower absorbance values. Once the absorbance versus concentration relationship has been established by a calibration curve through the plotting of absorbance values measured for standard samples at different concentrations, the unknown sample concentration can be determined when its IR absorbance is measured. Sample concentration can be calculated as

$$C = \frac{A}{eL} \tag{8.3}$$

A field air sample will likely contain more than one chemical that will absorb IR radiation. Simple application of Beer's Law can be problematic since different chemicals absorb IR of same frequency differently. The total absorbance (A_T) of a certain frequency (v_1) is the sum of the absorbance of each chemical (C_1, C_2 ... C_n) for that frequency:

$$A_{T,v_1} = A_{C_1,v_1} + A_{C_2,v_1} + \cdots + A_{C_n v_1} \tag{8.4}$$

For multiple wavelengths, a set of similar equations can be obtained. These equations can be resolved using matrix algebra, and the concentration of each chemical can be determined.

It should be understood that for IR spectrometers, Beer's Law might not be followed perfectly, especially at the low end of detectable limits. Therefore, field detection using IR is generally considered qualitative with semiquantitative features. For quantitative analysis of a sample, calibration curves using high-quality reference materials are required to determine sample concentration.

8.1.4 IR Transducers

After the IR passes through the sample, the resulting IR radiation is monitored using a transducer. An IR transducer changes the received IR signals to electrical signals that can be processed. Thermal, pyroelectric, and photoconducting transducers are common.

Thermal transducers are based on the following principles: (1) some materials build up electrical potential when there is a temperature difference (thermocouples); and (2) others change resistance when temperature changes (bolometer and thermistor). When a thermal transducer receives the IR radiation, its temperature changes, and therefore its potential or resistance changes as a consequence. These changes generate electrical signals that are then processed.

Simply put, a pyroelectric transducer is a piece of pyroelectric material placed between two electrodes. Pyroelectric material produces electrical polarization, such as in a crystal, via a change of temperature. The charge distribution on the pyroelectric material changes when it senses temperature variation. Therefore, the received IR at different intensities will produce electric signals that can be processed.

Photoconducting transducers use a thin film of semiconductor material. The conductivity of the thin film alters when it receives IR radiation. This change of conductivity is then detected and processed.

IR transducers are usually enclosed in a vacuum chamber to protect them from environmental variations, such as temperature fluctuations.

8.1.5 Photoacoustic Effect

The photoacoustic effect, which transforms electromagnetic radiation into acoustic waves, is also used as an IR detection technique in both qualitative and quantitative analyses of various materials. Alexander Graham Bell (1847–1922) discovered the effect in 1880. He observed that when a beam of sunlight was interrupted rapidly, sound was produced from the diaphragm he used in his experiment.

The photoacoustic effect is the generation of acoustic waves as a consequence of light absorption. As previously discussed, a molecule is excited to a higher vibration state when it absorbs energy from IR radiation. The higher energy of the excited molecule releases the absorbed energy in radiative decay and nonradiative decay. Radiative decay is the process of emitting radiation spontaneously, and nonradiative decay is the transfer of energy into mechanical vibrations of neighboring atoms. As a consequence, both types of decay raise the temperature of the sample. This increase in temperature increases pressure, and creates pressure waves in the gas. Pulsing, chopping, or modulating IR radiation at a certain frequency — for example, at a frequency of an audible sound wave — will produce a pressure wave

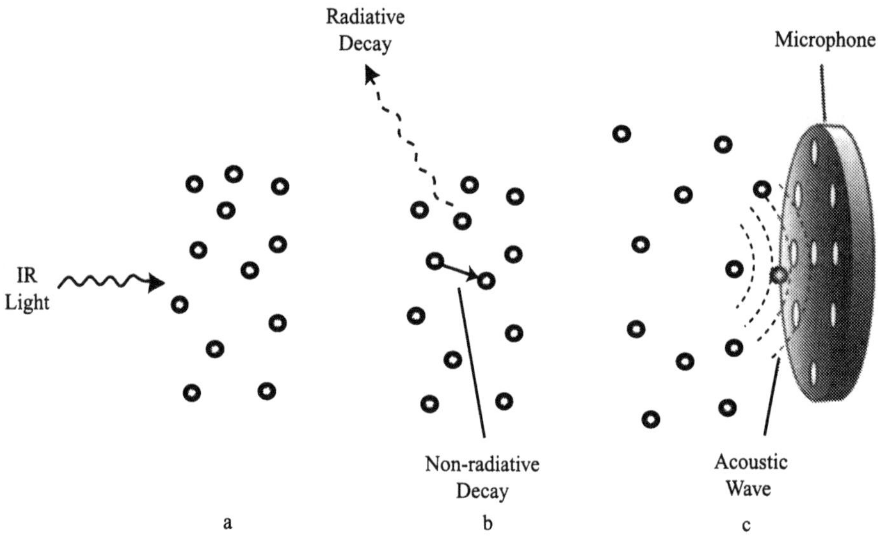

Figure 8.4 Photoacoustic effect principle.

at the same frequency of the modulation and generate an acoustic wave. If the gas is enclosed in a cell with a microphone or other measuring device such as an attached piezoelectric device, the pressure waves generated can be detected.

In a detector based on the photoacoustic effect (photoacoustic spectroscopy), an IR beam at a known frequency is used as the irradiation source. The IR radiation used should have a frequency at which it can be absorbed by the targeted chemical. This IR radiation is modulated at a desired frequency. Exposing the sample in a closed cell to this modulated IR frequency will generate the characteristic acoustic wave if the sample contains the targeted chemical. Signal amplitude is directly related to sample concentration. A schematic diagram illustrating the photoacoustic effect principle is shown in Figure 8.4.

8.2 INSTRUMENTATION OF FILTER IR AND PHOTOACOUSTIC DETECTORS

Many different kinds infrared instruments of various sizes are available for quantitative and qualitative vapor analyses. Some are manufactured specifically for detection of only one chemical, such as the carbon monoxide (CO), using a specific wavelength. Others can be very complicated, such as Fourier transform–based infrared (FT-IR) detectors that scan the entire IR wavelength for both chemical identification and concentration determination.

The absorption spectrum of CWAs or TICs is very likely known, or can be obtained in the laboratory and incorporated into libraries. Therefore, it may not be necessary to scan through the entire IR absorption spectrum of the sample. Exposure of the sample to a few important wavelengths that correspond to significant absorption wavelengths of the targeted chemicals may be a simpler and effective way for manufacturing point-sampling detectors. Since wavelengths are filtered so that only desired IR wavelengths reach the sample, detectors based on this method are called filter IR detectors. The vapor sample in a cell is bombarded with one or several known wavelengths from an IR source. Detection of changes in IR absorbance by the sample with a transducer can be related to determine sample concentration. If the signal is generated based on photoacoustic effect, the instrument is a photoacoustic detector.

8.2.1 IR Source and Wavelength Control

Among the most common ways of generating IR radiation is a solid block of rare-earth element oxide or silicon carbide electrically heated to a temperature of approximately 2000 K. A wavelength selector, either a monochromator or a wavelength filter, is used to achieve desired wavelengths from the IR source since the source produces radiation with a broad range of wavelengths. A monochromator is a device that separates radiation into its distinct wavebands using a diffraction prism. A wavelength filter is a device that allows only the radiation of a certain wavelength to pass through while suppressing radiation in other wavelengths. The sample could be irradiated with more than one of the characteristic wavelengths to achieve more reliable identification of the targeted molecules.

8.2.2 Sample Cell

The sample is drawn into or continuously through a sample cell of fixed dimensions. IR radiation enters the sample cell through a window made of material that permits the IR beam to penetrate. This cell and window are set in such a way that the window is perpendicular to the radiation path to permit maximum IR radiation. The IR beam can pass through the cell and reach the absorption photometer in a single pass, or in multiple passes through the use of properly aligned mirrors to reflect the IR beam multiple times back and forth through the sample cell to increase the effective path length. Highest sensitivity is achieved when path length is maximized.

8.2.3 IR Intensity Detectors

The absorption of IR can be detected by using one of the IR transducers or photoacoustic methods. Figure 8.5 illustrates how transducers measure the intensity change of IR absorption after it passes through the sample. The absorbance of the sample can be calculated based on IR radiation beam intensity (Equation 8.1). Since absorbance of the IR follows Beer's law and the path length of the cell is a constant, substance concentration can be calculated using Equation 8.3.

8.2.4 Photoacoustic Detectors

As discussed earlier, the photoacoustic effect can be used in IR instruments as a way to detect the presence of targeted substances. A diagram illustrating the operational principle of photoacoustic spectroscopy is shown in Figure 8.6. After an

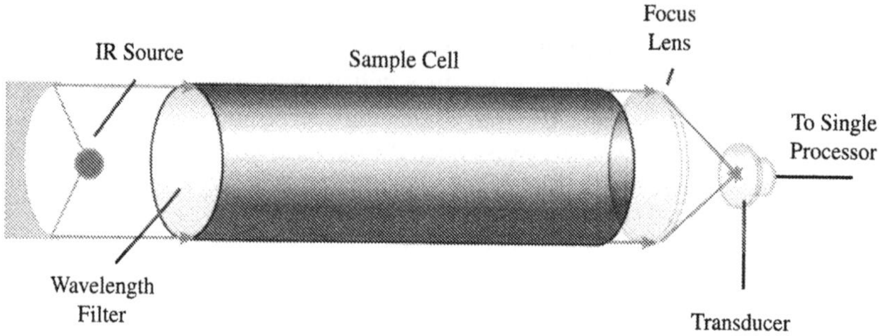

Figure 8.5 Schematic diagram of signal wavelength IR detector.

IR beam passes through the wavelength selector, the beam is modulated with a radiation chopper to a desired frequency. This can also be achieved by pulsing the IR radiation. This modulated radiation then enters the sample cell through the window. If the substance that absorbs the IR at the selected wavelength is present in the vapor, a photoacoustic wave will be generated. Through either a highly sensitive microphone or a piezoelectric sensor or other detector such as pressure sensor, the acoustic signal is translated into an electrical signal that can be further processed. The magnitude of the pressure generated by the expanding gas is proportional to the concentration of the substance that absorbs the IR radiation. Consequently, concentration is determined based on the magnitude of the measured signal.

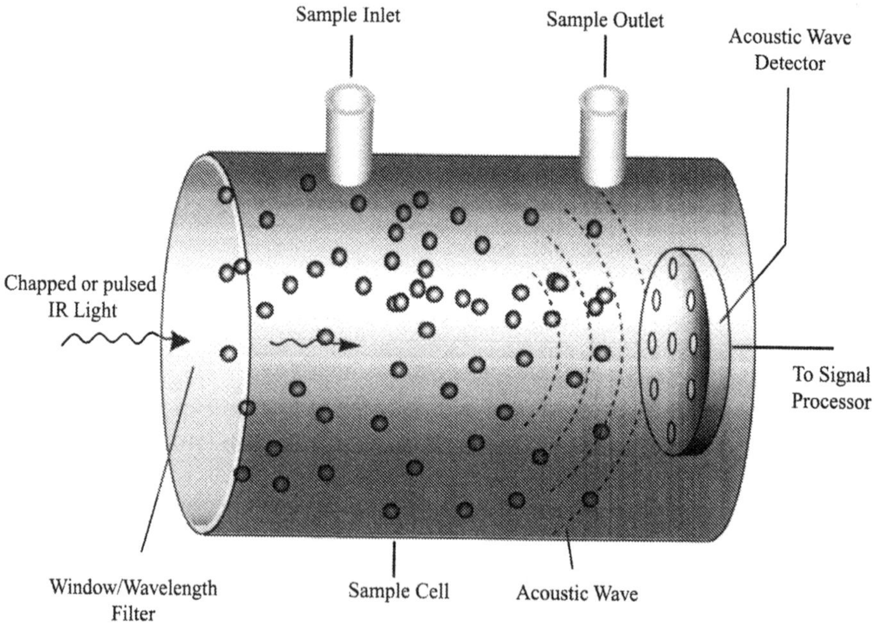

Figure 8.6 Schematic diagram of photoacoustic spectroscopy.

8.2.5 Detector Operation

The filter IR detector, similar to most other spectrophotometers, consists of an IR radiation source, specific wavelength filter, sample cell, transducer, and signal processor. Vapor is drawn into the sample cell by an internal pump through the sample inlet and irradiated by an IR beam through the filter. Analysis may begin after the sample vapor fills the sample cell. Usually a parabolic mirror is used to focus the radiation beam. If the sample contains the targeted substance, the transducer detects it in the intensity of the exiting beam. For instruments equipped with a monochromator that varies wavelengths automatically or manually, changing filters is not necessary. The monochromator permits automatic selection of multiple wavelengths of interest to penetrate the sample such that various function groups can be detected simultaneously.

Similarly, photoacoustic IR detectors consist of an IR source, wavelength filter, radiation chopper, sample cell, microphone or piezoelectric detector, and signal processor. The chopper is used to modulate the IR radiation before it enters the sample cell, which in turn occurs after the sample cell is filled and closed. Here the sample is not continually introduced through the cell; instead the sampling procedure is pulsed for each analysis. IR radiation, modulated at a certain frequency, is used to excite the vapor sample. The generated photoacoustic effect signal is captured by the signal sensor and sent to the processor.

Both types of detectors do not analyze sample continuously. Each sample is pumped into the detector for analysis, and the sample is purged after each analysis and the cell is flushed with clean air. Analysis resumes after a new sample is introduced into the cell. Similar to photoionization techniques, discussed in Chapter 11, this technique is nondestructive. The chemicals in the sample are not decomposed or altered chemically by the analysis.

8.2.6 Technique Specification of Filter and Photoacoustic IR Detectors

8.2.6.1 Detectable Substances and Selectivity

Since CWAs and many TICs absorb IR of certain wavelengths, they can be, at least theoretically, detected using IR detectors. Selectivity is controlled by the careful selection of wavelengths for each of the targeted chemicals. To enhance selectivity, a sample may be exposed to IR radiation of several characteristic wavelengths. Various chemicals may contain the same function groups that absorb IR of similar wavelengths. Detection determination based on a single wavelength may result in higher false positive and/or false negative alarm rates due to reduced discriminatory ability. When the detected sample produces an alarm suggesting the presence of a targeted chemical, checking the sample using one or more additional characteristic wavelengths to confirm the finding is recommended. This is possible because the IR method does not destroy sample integrity. After exposure to IR radiation of a certain frequency, the sample in the cell is chemically unchanged and can be analyzed again. This technique improves sample identification reliability. To do this, after the sample in the cell is analyzed with a single frequency, it is then exposed to a second

IR radiation at another frequency to supplement or confirm the detection observed in the first analysis. A third wavelength can be used, if necessary, to further enhance detection reliability.

Unlike other techniques in which samples can be continually introduced into the detector, these two techniques use pulsed sampling methodology. The sample is introduced into the sample cell, which is then closed during the analysis. Sample volume can be rather large. Therefore, no GC separation technique is coupled with these types of detectors currently.

8.2.6.2 Sensitivity and LOD

Both IR techniques discussed above are believed to have relatively high sensitivity with low LODs. Under ideal conditions, IR detectors may be able to sense CWA concentrations at the lower parts-per-million level.

8.2.6.3 Environmental Effect

Several factors can affect IR detector performance. Relative humidity is among the most notable. Moisture molecules composed of two hydrogen atoms and an oxygen atom is an asymmetric molecule that strongly absorbs IR radiation. Changes in sample moisture level may generate false IR absorption responses, as well as interfering with the calibration. To reduce this effect, a filter for measuring background moisture content can be used, which then facilitates calculating a correction.

The photoacoustic technique is very sensitive to vibration and environmental effects. Thus, its use in handheld detection devices for field operations is limited at present.

8.2.6.4 Response Time

The analysis cycle includes sampling time, IR analysis time, and recovery time. Among them, the sample introduction and cell flushing are main factors affecting an analysis cycle. The actual IR analysis response time can be within seconds. Total cycle for a complete analysis can be several seconds to minutes and depends on the size of the cell and the rate of sampling.

8.3 FOURIER TRANSFORM INFRARED DETECTORS

The Fourier transform infrared (FT-IR) spectrometer scans a large range of IR frequencies by using an interferometer, while simultaneously acquiring the spectra. Analysis with an FT-IR spectrometer actually includes two main steps. The instrument collects the interferogram of the sample using the interferometer and then applies the Fourier transform to the interferogram to obtain the final IR spectra. FT-IR spectrometers once were very expensive. However, thanks to development of related technologies, prices have dropped significantly and small detectors can be made based on this technique.

Detectors based on dispersive IR techniques also scan a large region of IR and obtain the spectra. The dispersive IR technique employs a monochromator or other methods to generate IR radiation of various wavelengths. It has to sequentially produce different wavelengths in the IR region of interest. This process can be slow with low resolution due to limitations on controlling wavelength. The FT-IR spectrometer significantly improves analysis time, resolution, and other performance-related issues. FT-IR has also largely replaced dispersive IR. Because FT-IR scans broadband IR in a very short time, it enables the IR technique to be used along with a GC or high-performance liquid chromatographic separation. FT-IR spectrometers have also been used for detecting CWAs and TICs.

8.3.1 Interferometer

The heart of an FT-IR spectrometer is the interferometer. Most FT-IR spectrometers use the Michelson interferometer, invented by Albert A. Michelson (1852–1931) in 1891, and for which he received the 1907 Nobel Prize in physics.

Basically, the Michelson interferometer, illustrated in Figure 8.7, is composed of a beam splitter, fixed mirror, and moving mirror. Incoming IR radiation is split into two beams that are reflected by the mirrors and then recombined before reaching the IR transducer. Changing the moving mirror position enables the interference of the two beams to form an interferogram that is used to obtain the spectra.

As Figure 8.7a shows, the IR beam (1) emitted from the IR source strikes the splitter, which then divides the beam into two vertical beams (2 and 3). Due to the special property of the splitter, about 50% of the incoming beam is reflected by the splitter toward the fixed mirror and the other 50% is allowed to penetrate the splitter to hit the moving mirror. These beams are reflected back by the fixed and moving mirrors toward the splitter (Figure 8.7b). The splitter then reflects half of the beam

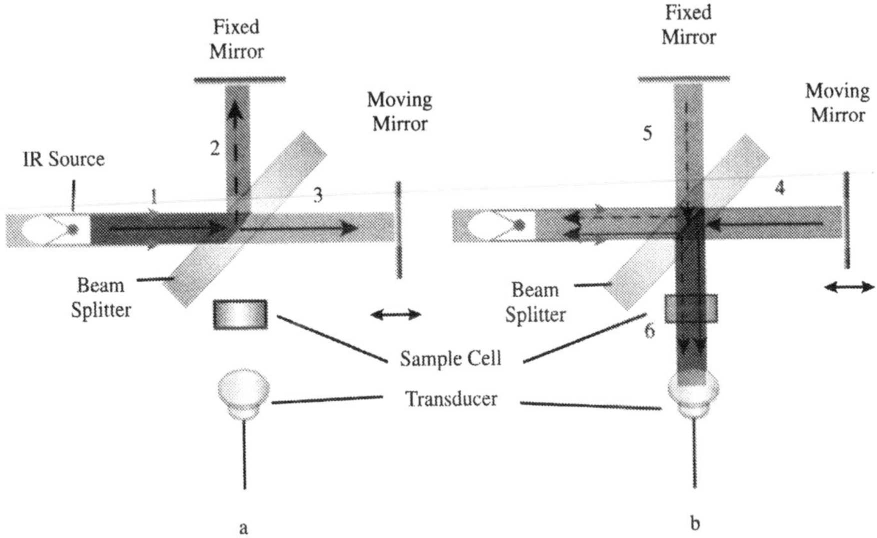

Figure 8.7 Schematic diagram of Michelson interferometer.

from the moving mirror (4) to the transducer, and also allows the other half of the beam from the fixed mirror (5) to penetrate it. These two beams then recombine (6) and pass through the sample cell. The rest of the radiation that is reflected from the moving mirror penetrates the beam splitter in the direction of the radiation source. Similarly, the rest of the radiation from the fixed mirror is reflected by the splitter in the direction of the radiation source. Therefore, only about half of the initial IR radiation goes through the sample cell and reaches the transducer.

The recombination of the beams after the splitter and the sample cell results in the interferogram, which is a plot of the output power versus the difference (δ) in path lengths for the two recombined beams. The δ value depends on the relative position of the moving mirror. Since the radiation strikes the moving mirror and is then reflected back, δ is equal to two times the difference between the path lengths of the moving mirror (M) and the fixed mirror (F) (Figure 8.8a). For radiation of a single frequency, the output power will be maximum when the two beams that recombine are in phase at $\delta = 0$ (Figure 8.8b). This occurs when the moving mirror is at position (a). When the moving mirror moves, the power of the recombined beam changes due to the destructive interference of the two beams. Figure 8.8c illustrates the recombination of the beams when the moving mirror moves a distance of $\lambda/8$ ($\delta = \lambda/4$) to position (b). When the moving mirror moves $\lambda/4$ ($\delta = \lambda/2$) to position (c), the two beams will be totally out of phase, and the lowest output power is observed (Figure 8.8d). As the IR radiation is composed of different wavelengths, the output power will vary when the moving mirror is at different positions and an interferogram is observed. If the sample in the sample cell absorbs the incoming IR of certain wavelengths, a characteristic interferogram is generated. By applying the Fourier transform to the interferogram, a distinctive spectrum is then obtained.

8.3.2 Fourier Transform

Discussion of how the Fourier transform is applied to the interferogram is beyond the scope of this book. Suffice it to say that data are obtained for each precise position of the moving mirror in small intervals during the scan to acquire an interferogram. This interval determines the resolution of the FT-IR. In an FT-IR spectrometer, a pyroelectric transducer is commonly used due to its fast response to signal changes. The observed interferogram is digitized and stored in the computer. Since the interferogram contains information about every frequency in its IR source, individual frequency intensity is therefore found by applying the fast Fourier transform to construct the spectra.

8.3.3 Background Handling

A background spectrum that is converted from a background interferogram by Fourier transform is usually first obtained for correcting effects from common factors such as water, CO_2, and noise from the IR source, interferometer, and detector, among other sources. The sample spectrum is then obtained. The difference between the sample and background spectra gives the final spectrum that is corrected for effects from common factors. This process enables more precise identification of unknown substances.

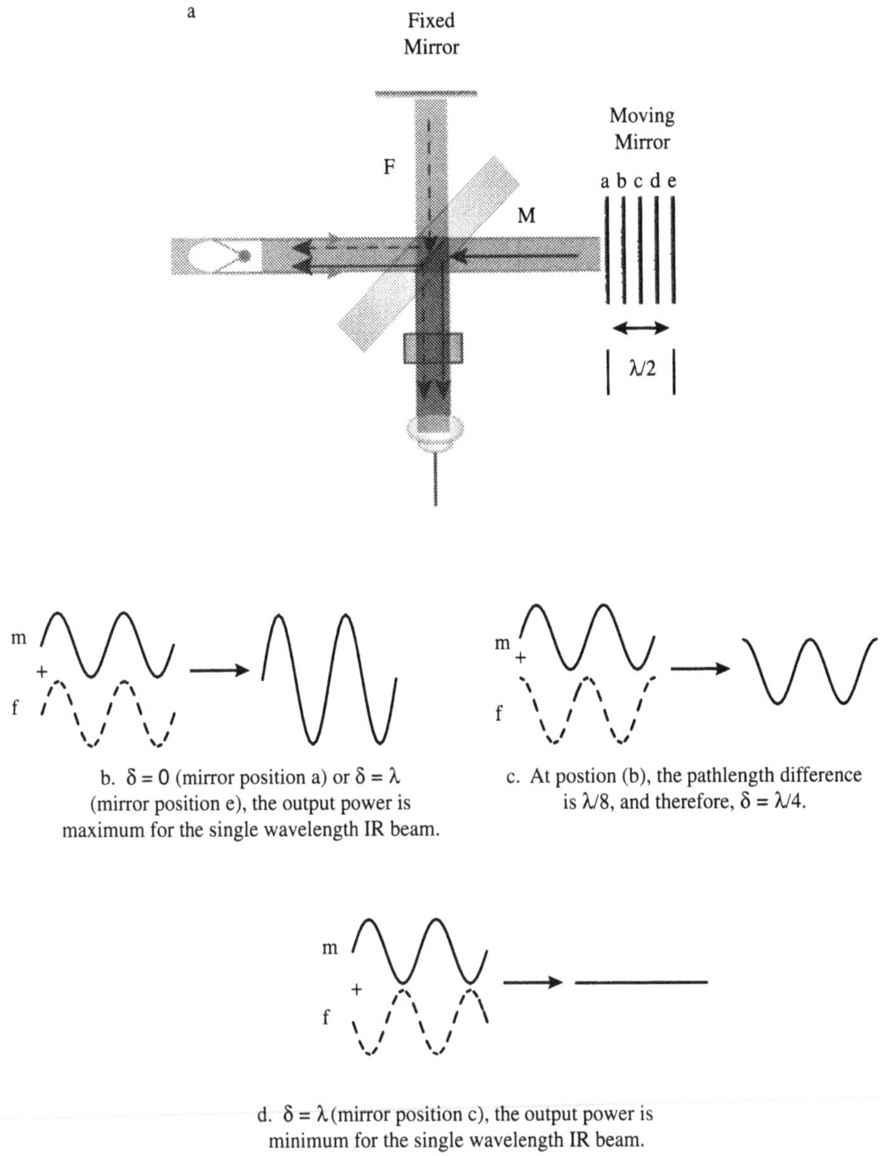

Figure 8.8 Formation of interferogram.

8.3.4 CWA and TIC Detection

FT-IR can be used for detection of CWAs and TICs in vapor, liquid, or solid samples. The most common method used in identifying chemicals in samples is to compare the resulting spectrum to the various libraries stored in the instrument. For instruments used in detection of CWAs and some TICs, the library may be limited to the chemicals of interest for faster search results. This is different from attempts to identify unknown chemicals, as the spectrum may have to be compared with a large database for positive identification. Concentration can be determined as discussed in Section 8.1.3.

8.4 REMOTE IR MONITORS

The spectrometers discussed thus far use an internal IR source to generate desirable IR radiation for detection. Since IR radiation is part of the environment, it is possible to monitor changes in background IR radiation. Remote IR monitors work in a similar fashion to human eyes, except that the monitors sense IR radiation instead of visible radiation. The presence of CWA vapor in the background will change the incoming IR spectra. By comparing incoming spectra with stored library patterns, the detector can determine whether a CWA cloud exists within a certain range. As it scans the area, the monitor could also report the position of the cloud and direction of its movement.

8.5 APPLICATIONS

Laboratory instruments can provide accurate identification of various substances through library searches of spectra collected with known substances. Detection devices using IR techniques in field applications include remote and point sample detectors. The M21 Detector and the Rapid Air Pollution Infrared Detector (RAPID® or HAWK®), used by the U.S. military in field operations, are examples of remote devices that detect IR radiation changes in the background. The MIRAN SapphIRe Portable Ambient Air Analyzer (Figure 8.9) manufactured by Foxboro Company is an example of how IR technology is used in a portable instrument. The MIRAN SapphIRe has been tested at the U.S. Army Soldier and Biological Chemical Command (SBCCOM) under the Domestic Preparedness Program in an attempt to qualify it for possible CWA detection.

Figure 8.9 MIRAN Detector. Photo courtesy of K. Y. Ong.

The MIRAN SapphIRe portable ambient air analyzer is an IR spectrophotometer. The intensity of the IR energy absorbed by the air sample in the cell is displayed on the detector panel in absorbance units (AU). The optimum detection wavelength for the compound of interest must be stored in the instrument's memory to enable detection of that compound. The factory usually provides the appropriate wavelength to be used for detection of a specific substance. The instrument also has a multigas mode that allows the detector to indicate absorption responses at three wavelengths simultaneously. The MIRAN SapphIRe can operate at different path lengths. The longest path length is 12.5 m, which provides the highest possible sensitivity of the instrument toward detection of CWA vapors. It is equipped with an external particulate filter to prevent dirt from entering its internal plumbing, and a chemical filter that is interchangeable with the particulate filter to zero the instrument as well as to provide clean air for the instrument in a contaminated environment.

The Multi-gas Monitors® 1312 and 1314, manufactured by Innova Airtech Instruments A/S (Denmark), are examples of instruments using the photoacoustic technique. Another type of detector uses FT-IR technology to analyze liquid or solid samples. The Travel*IR*® (Figure 8.10) manufactured by SensIR Corp. has also been tested at SBCCOM under the Domestic Preparedness Program for its application in identifying unknown samples including those that may contain CWAs. The Travel*IR* is an FT-IR spectrometer that can be used to identify unknown solids, powders, pastes, gels, and liquids. The technique involves placing a sample on top of a diamond crystal (which has a high refractive index) embedded in a stainless steel disk, called the DuraDisk®. An infrared beam from the spectrometer passes through the crystal and penetrates slightly into the sample to permit analysis of strong infrared absorbing solutions, such

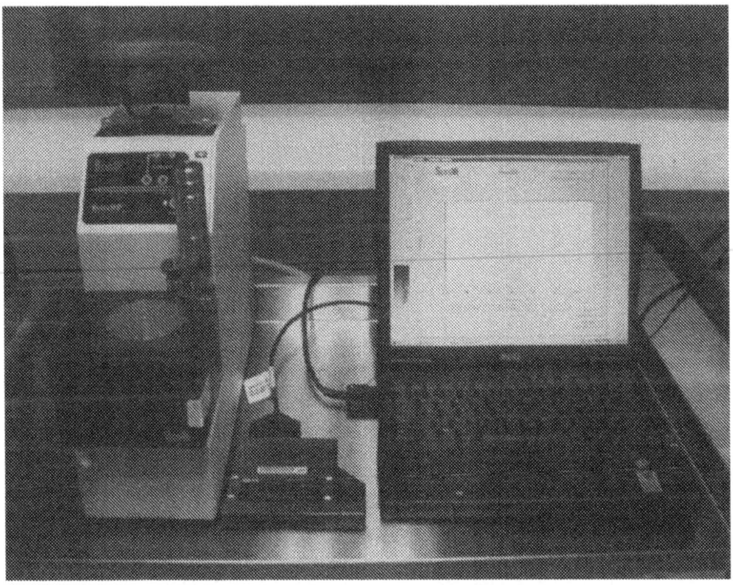

Figure 8.10 Travel*IR* Detector. Photo courtesy of K. Y. Ong.

as emulsions or aqueous solutions. The technique can also be useful for solid samples. The process is nondestructive of the sample and the instrument.

Unknown samples are identified by looking for pattern matches in reference database libraries. The Travel*IR* can identify pure substances readily and reliably with reproducibility. However, identification of a substance in a mixture is considerably more difficult. The complex spectrum from the mixture often leads to erroneous results. For example, a mixture of the nerve agent GB and diesel fuel showed a higher spectral quality value for GF (another nerve agent) than for GB itself. Perhaps this was due to the structural and spectral similarity of these two agents and incomplete compensation in the C-H (carbon-hydrogen) stretching region. The problem becomes much more complex when the CWA is dissolved in or interspersed with different substances. Therefore, the Travel*IR* should be used in conjunction with other approved CWA detectors. If the solvent containing the agent is volatile and its evaporation leaves behind a concentrated sample of CWA, the Travel*IR* can readily identify the agent. These tests led to a chloroform extraction technique tried on VX mixtures. These findings were encouraging. Perhaps, by implementation of a simple extraction process, the Travel*IR* will be much more useful for identifying potential CWAs in a mixture.

The Transport Kit® (Figure 8.11) from Thermo Electron Corp. is similar to the Travel*IR*. Testing of the Transport Kit is underway as of this writing.

Results of testing the Travel*IR* could provide insights leading to successful application of this FT-IR technology in identifying air samples. The Travel*IR* has the software capability to subtract the suspected background spectrum to isolate the substance spectrum for subsequent identification search. Such a feature may provide a means of minimizing potential interference.

Figure 8.11 Transport*IR*100 Detector. Photo courtesy of K. Y. Ong.

Unfortunately, the tested detectors based on IR techniques do not have sufficient sensitivity to detect CWAs in vapor to meet IDLH levels. Given the ability of these instruments to reliably identify targeted chemicals, they could become useful tools in assessing incidents where concentrations higher than IDLH levels are likely to exist.

8.6 FACT SHEETS FOR SELECTED IR-BASED DETECTORS

Detector Name

HAWK Long-Range Chemical Detector

Manufacturer/Distributor and Contact Information

Bruker Daltonics Inc.
40 Manning Park
Billerica, MA 01821
Tel: 978-663-3600

Technique Description

The HAWK is a portable instrument that uses infrared for stand-off remote sensing detection of a chemical cloud. It is a passive IR detector that identifies the characteristic signatures of chemicals in the near-infrared spectrum. Identification is made through spectrum matching from libraries.

Chemical Detection Capability and Performance

The detector is designed to serve as a remote sensor for the presence of CWAs and important TICs. The sensor is mounted on a tripod or on vehicles that permit 360° field of regard coverage at an angle of −10° to 50° using a spatial scanner in 3 sec. It has a spectral range of 700 to 1300cm^{-1} and a scanning rate of 20 spectra/sec at 4 cm^{-1} resolution.

Other Features

Includes visual indication of alarm through connection with a PC via the RS 422 interface using Windows-based interface displays. Audio and remote alarms are also available. Interface for network connections. The PC software graphically visualizes the detected cloud's extension and direction. Built for rugged environment operations.

Detector Name

HazMatID™/Travel*IR*-HCI™ Chemical Identifier

Manufacturer/Distributor and Contact Information

SensIR Technologies
14 Commerce Drive
Danbury, CT 06810
Tel: 203-207-9700

Technique Description

These portable instruments use infrared detection for solid or liquid chemical identification via identifying characteristic chemical signatures. Identification is done through spectrum matching from many libraries. A spectra subtraction routine permits identification of components in a matrix.

Chemical Detection Capability and Performance

The detector is designed to serve as an on-the-spot sensor for identification of unknown solid or liquid substances for the existence of CWAs, drugs, bioagents, and TICs, all within a few minutes. The HazMat ID is the waterproof version of the Travel*IR* that can identify simple mixtures through automated subtraction software. The unit can operate in hostile environments and can be immersed in decontamination solution without harm to the internal mechanism.

Other Features

Visual component identification occurs through connection with a PC or built-in Windows-based interface displays. Interface for network connection. PC software graphically visualizes the spectra for comparison.

Detector Name

Innova Photoacoustic Infrared Spectroscopy

Manufacturer/Distributor and Contact Information

Innova Airtech Instruments A/S (Denmark)
Distributor:
California Analytical Instruments Inc.
1238 West Grove Avenue
Orange, CA 92865-4134
Tel: 714-974-5560

Technique Description

These instruments comprise IR detectors using photoacoustic infrared spectroscopy for monitoring the ambient air for targeted chemical(s). Photoacoustic audible signals created from bombardment of the sample with modulated light are detected. Chemical agents are distinguished from interference chemicals by relative signals when sequentially bombarded by modulated light at different wavelengths. Interference rejection improves when more wavelengths are used to identify an unknown.

Chemical Detection Capability and Performance

The detector is designed to serve as an on-the-spot monitor for the presence of CWAs, TICs, or other chemicals within seconds. The instrument is sensitive to external vibration and humidity effects, and requires calibration in each operating environment immediately before use.

Other Features

Visual indication of component spectrum. The instrument has not been tested under the Domestic Preparedness Program.

Detector Name

JSLSCAD Joint Service Lightweight Stand-off Chemical Agent Detector

Manufacturer/Distributor and Contact Information

General Dynamics
128 Lakeside Ave.
Burlington, VT 05401
Tel: 703-414-1800 (Crystal City office)

Technique Description

This instrument is a fixed-site detector using infrared for stand-off detection of a chemical cloud. It is a passive IR detector that identifies the characteristic signatures of chemicals in the near-infrared spectrum. Identification is through spectrum matching from libraries.

Chemical Detection Capability and Performance

The detector is designed to serve as a remote sensor for the presence of CWAs and important TICs. The sensor is mounted on various platforms for ground, aerial, and shipboard operations that permit a 360° field of regard coverage at an angle of –10° to 50° using a spatial scanner. It searches the 7 to 14 micron region of the surrounding atmosphere for chemical agent vapor clouds.

Other Features

Visual alarm indicator through RS232/422 interface using Windows-based displays. Audio and remote alarms are also available. Interface for network connections. PC software graphically visualizes the detected cloud's extension and direction up to 5 km. Built for rugged environment operations.

Detector Name

MIRAN SapphIRe Portable Ambient Air Monitors Model (1, 5, 30, 30E, 100, and 100E)

Manufacturer/Distributor and Contact Information

The Foxboro Company
33 Commercial Street
Foxboro, MA 02035-2099
Tel: 888-369-2676

Technique Description

These instruments are single-person portable, single-beam IR detectors used for monitoring the ambient air for the presence of targeted chemical(s). The number of chemicals (stored in the internal library) that can be monitored is indicated by the model number. The high-numbered models offer additional features. The E series permits the user to avoid interference by nontargeted chemicals by selecting a peak wavelength that effectively filters them out.

Chemical Detection Capability and Performance

This device is designed to serve as a near real-time monitor for CWAs, TICs, or other chemicals. The instrument displays the detected chemical concentration in parts per million, parts per billion, percentage, milligrams per square meter, or absorbance units. The instrument draws in 15 L of sample per minute to fill the gas cell volume of 2.23 L. The 100E has seven fixed bandpass filters and one continuously variable filter (7.7 to 14.1 microns) together with two selectable path lengths (0.5 and 12.5 m).

Other Features

Visual indication of component concentrations. Operable in temperature range of 5°C to 40°C at noncondensing relative humidity. Serial port for linking with PC.

Domestic Preparedness Test Report

Domestic Preparedness Program: Testing of MIRAN SapphIRe Portable Ambient Air Analyzer [100E] Against Chemical Warfare Agents Summary Report, July 2000 (http://hld.sbccom.army.mil/ip/reports.htm#detectors).

Detector Name

Remote Sensing Chemical Agent Alarm, M21 (RSCAA4)

Manufacturer/Distributor and Contact Information

Tradeways Ltd.
184 Duke of Gloucester Street
Annapolis, MD 21401
Tel: 410-295-0821

Technique Description

This instrument is a stand-off type of CWA detector that uses passive IR for remote detection. It provides early warning of CWAs in vapor form, including GA, GB, GD, HD, and L from distances up to 5 km.

Chemical Detection Capability and Performance

It can detect nerve and blister agent surface contamination at approximately 90 mg/m^2 of nerve and 500 mg/m^2 of HD. Interference is eliminated by using the FT-IR detection/discrimination algorithm at 1250 to 780 cm^{-1} region.

Other Features

Audio and visual alarm indicators. Can be upgraded to detect new compounds. Operable in a field environment between –25°C to 48°C, and met military specifications.

Detector Name

Transport IR-100 Spectrophotometer Chemical Identifier

Manufacturer/Distributor and Contact Information

Thermo Nicolet Corporation
5225 Verona Road
Madison, WI 53711-4495
Tel: 800-642-6538

Technique Description

This portable instrument uses infrared detection for solid or liquid chemical identification, based on characteristic chemical signatures. Identification is through spectrum matching from many libraries. A spectra subtraction routine permits identification of components in a matrix.

Chemical Detection Capability and Performance

This detector is designed to serve as an on-the-spot sensor for identifying CWAs, drugs, bioagents, and TICs in solids or liquids within few minutes. The IR100 has a built-in computer, or can be connected to a PC. It can identify simple mixtures through the EZ OMNIC software.

Other Features

Visual indication of component identification through connection with a PC or built-in displays. Interface for network connection. The PC software graphically displays spectra for comparison. This instrument has not been tested under the Domestic Preparedness Program as of this writing.

CHAPTER 9

Surface Acoustic Wave and Electrochemical Techniques

Surface acoustic wave (SAW) technology utilizes the production, control, transmission, and reception of acoustic waves formed on the surface of a piezoelectric substance. It has been used in telecommunications applications for over 50 years. SAW devices are used in mobile cellular phones and base stations as band-pass filters, and also serve as sensors for many other applications such as torque, pressure, temperature, humidity, and chemical sensors.

The first chemical agent sensor using SAW technology was reported in 1979 by Wohltjen and Dessy. Currently, many variations of SAW-based chemical sensors are under development around the world. For example, the current Joint Chemical Agent Detector (JCAD), under development by the U.S. military, uses SAW technology. SAW devices as viable future CWA detectors are being aggressively developed. The race is on to produce detection devices that meet JCAD requirements.

The heart of a SAW device is a piezoelectric crystal plate. When an electrical field is applied to the crystal, a surface acoustic wave is generated and propagates on the surface of the piezoelectric substrate. Any changes in characteristics of the propagation path affect the wave frequency. For example, additional deposits of substances on a surface change the surface characteristics, and thus affect the wave propagation frequency. This change in frequency is monitored and can be correlated to the corresponding quantity of deposit being measured. Consequently, the amount of CWA loaded on the piezoelectric substrate surface upon exposure to a vapor sample can be detected.

SAW devices have some advantages over other technologies. The device can be manufactured at relatively low cost but with good detection sensitivity. It responds rapidly to chemicals deposited on its surface and can be miniaturized easily. Its reliability as a viable CWA detector is steadily being improved through aggressive research and development efforts.

Since both HAZMATCAD and JCAD employ electrochemical (EC) sensors to detect blood agents and other TICs, discussion of this topic is also included in this chapter.

9.1 PRINCIPLE OF OPERATION OF SAW DEVICES

Two major processes contribute to detection with a SAW device: generation and change of surface waves on a piezoelectric plate, and the sorption/desorption of chemicals on the surface.

9.1.1 Piezoelectric Effect

Figure 9.1 illustrates the piezoelectric effect. Piezoelectric plates are made of natural crystals (e.g., quartz) that do not possess center symmetry. When piezoelectric crystals are subjected to an external voltage, the crystals will expand or contract accordingly (Figure 9.1a). When the crystals are subjected to a mechanical strain (tension or compression), an electric voltage is generated (Figure 9.1b). This is called the piezoelectric effect, discovered by Pierre Curie (1859–1906) in 1880. When an AC field is applied at one end of the piezoelectric crystal plate, this end of the plate will expand and contract accordingly. This movement propagates through the crystal, or along the surface of the crystal, toward the other end of the plate. Electric voltage is generated by this expansion and contraction because of the piezoelectric effect (Figure 9.1c).

9.1.2 Surface Acoustic Waves

Applying an electric field to a piezoelectric crystal causes vibration resonation in several different ways, including flexure, lengthwise, area, radial, thickness-shear, thickness-trapped, and SAW, among others. Direction of vibration varies from one type to another.

Many chemical detection sensors are based on the SAW effect. In Figure 9.2, SAW occurs on a piezoelectric crystal plate with all the acoustic energy confined within one wavelength thickness of the surface. Since the wave propagates along the surface, the SAW sensor directly senses any mass change with associated changes in frequency and wave attenuation. SAW frequency may range from 25 MHz to 500 MHz or even higher. In general, frequencies greater than 50 MHz are used. Sensitivity is a function of the square of the vibration frequency; therefore, higher operational frequency enables increased sensitivity of the device to monitor surface perturbations. Because the wave propagates only along the surface of one side of the crystal, the other side can be bonded to a support base without interference in the detection process.

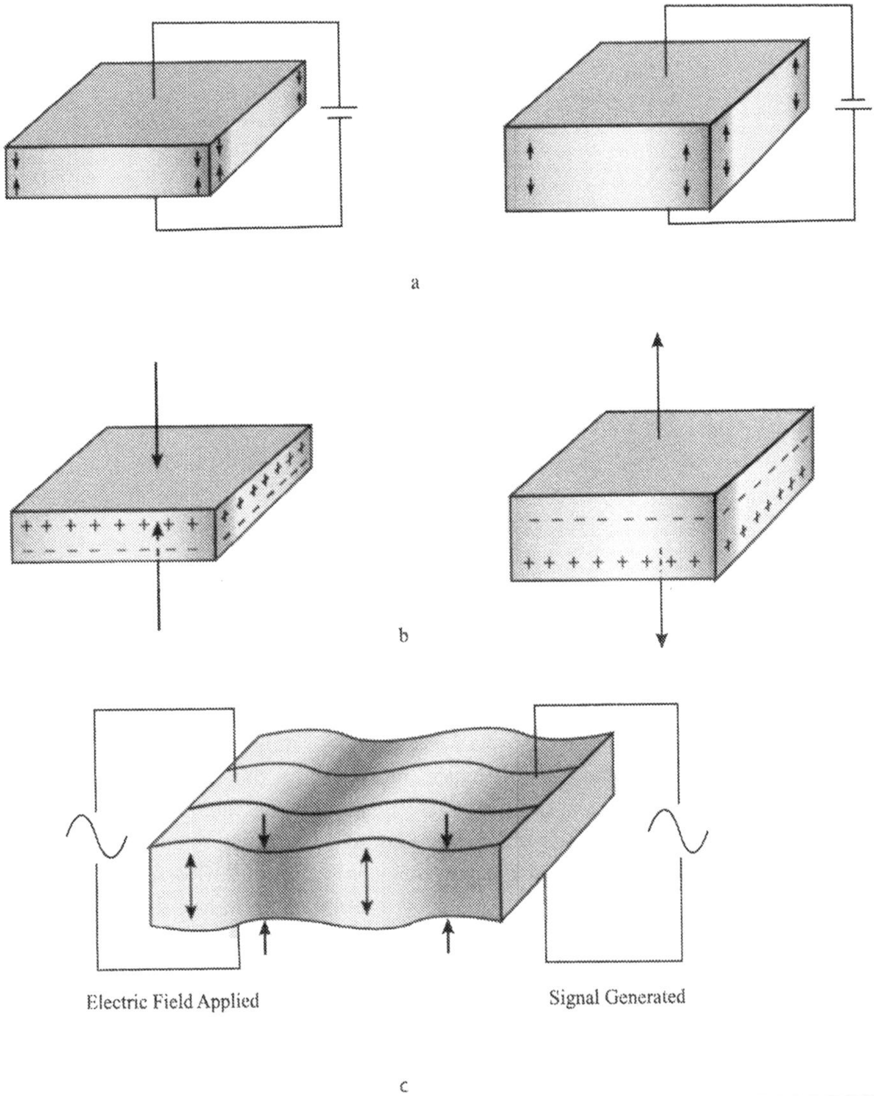

Figure 9.1 Piezoelectric effect.

9.1.3 Surface Sorption

To use the SAW device to detect chemicals, the targeted chemicals must be sorbed onto its surface to change the acoustic wave's propagation path property. The piezoelectric crystal does not itself have the desirable sorption ability to attract target chemicals. A thin layer of polymer substrate on the piezoelectric material is used to provide selective sorption of the targeted chemical from the vapor sample.

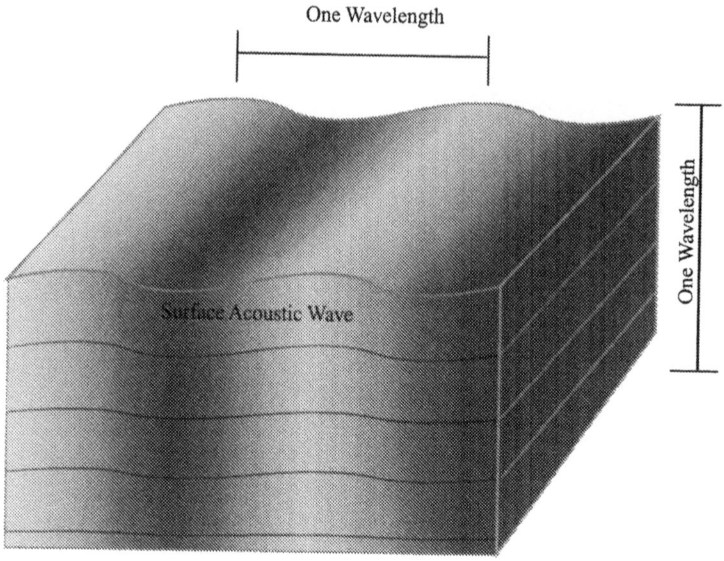

Figure 9.2 Surface acoustic wave.

A polymer can be comprised of a natural or a synthetic compound, consisting of millions of small, simple molecules called monomers. Each monomer is a link in the long-chained polymer molecule. Polymers have many free, active sorption sites that can effectively sorb the incoming chemical molecules (Figure 9.3). When a sample vapor enters the SAW detector, molecules in the vapor come in contact with the polymer surface at a certain rate depending on the vapor flow. When a chemical agent molecule hits the surface, it will be either bonded to the active sorption sites on the surface of the polymer (A in Figure 9.3) or deflected by the surface (B in Figure 9.3). The surface sorbs the molecules when they are captured on the active sites of the polymer. Adsorption occurs when the molecule (adsorbate) remains on the surface of the polymer (adsorbent).

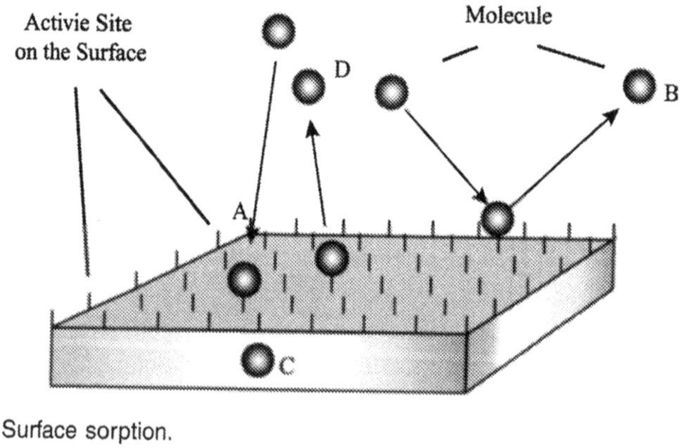

Figure 9.3 Surface sorption.

Absorption occurs when the molecule soaks into the coating rather than remaining on the surface (C in Figure 9.3). The difference between adsorption and absorption is adsorption occurs when the particle remains on the surface while absorption occurs when the substance enters the substrate. Sorption occurs when both happen simultaneously. Adsorption can reach equilibrium much faster than the absorption process. Consequently, the adsorption process requires a much shorter recovery time.

The simplest way to explain a sorption process is to describe it as a first-order reaction between the sorbate (S) and the sorbent (M):

$$M + S \xrightarrow{k_{ads}} MS \tag{9.1}$$

where k_{ads} is the sorption rate constant.

Disassociation occurs at the same time at a different rate:

$$M + S \xleftarrow{k_{dis}} MS \tag{9.2}$$

where k_{dis} is the desorption rate constant.

In the following, the sorption coefficient, K, represents the affinity between the sorbate and the sorbent, is the ratio of the two constants:

$$K = \frac{k_{ads}}{k_{dis}} \tag{9.3}$$

When adsorption equilibrium is reached, K equals

$$K = \frac{[MS]}{[M][S]} \tag{9.4}$$

The number of molecules that can be sorbed on the surface depends on several factors: the sorption coefficient (K) of the polymer to the chemical, the number of available sorption sites on the sensor's surface (M), and the concentration or vapor pressure of the sorbate in the vapor sample (S).

A high sorption coefficient means that the substance has strong affinity to the polymer, such that it can be sorbed to the sorbent effectively and rapidly. As expressed in Equation 9.4, a higher sorbed concentration (MS) at lower chemical concentrations yields a higher K value of the vapor at equilibrium. Chemicals with high K values will be sorbed more readily than those with low K values, as they compete for the available sorption sites. Therefore, chemicals with high K values are selectively sorbed on the polymer coating surface on the detector's SAW sensors.

The sorption coefficient decreases with temperature. The potential energy of a molecule is higher at higher temperatures. When the temperature reaches a certain level, the bond between the sorbent and the sorbate will break, resulting in release of the sorbed molecules from the polymer (D in Figure 9.3). This increases the dissociation coefficient (k_{dis}) and reduces the sorption coefficient (K). Therefore, the sorption ability of the polymer is reduced at higher temperatures. Effective release of the sorbed chemical from the SAW sensor surfaces can be induced with similar effect by heating the sensors to a higher temperature. This method is used to release the sorbed chemicals from the SAW sensors after each sample analysis.

The limit of detection (LOD) for various chemicals is determined by respective sorption coefficient values. The sensor polymer can attract more of the chemicals with higher K values from vapor onto its surface to produce a higher MS. Therefore, the given sensor is said to have a better limit of detection toward a compound that has a high sorption coefficient value. Thus, the detector can identify the higher K-value chemicals at lower concentrations.

The available active sorption sites M in a unit of surface area, and the vapor pressure or concentration of the chemical in the vapor that governs S are factors that determine how much adsorbate can be sorbed. Equation 9.4 can be rearranged to

$$[MS] = K[M][S] \tag{9.5}$$

Since K is a constant at a certain temperature, the increase of either M or S will increase the sorbed concentration on the surface, MS. For a SAW sensor, the surface area of the polymer coating is a constant. Thus, M can be considered as a constant also. Therefore, the responses of the detector are directly related to the concentration of the chemicals in the sample.

Equation 9.4 also suggests that decreasing the sample concentration in the vapor reduces the sorbed chemical on the surface. Therefore, the effective way to clean the SAW device after an analysis is the use of heat to reduce the sorption coefficient while using a clean carrier gas to flush the released sorbates out of the detector.

The above discussion is based on the establishment of sorption equilibrium. The equilibrium between the vapor and the surface is essential for quantitative analysis. Unless sorption equilibrium is reached, it is difficult, if not impossible, to determine vapor concentration. Fortunately, such sorption equilibrium can be reached within seconds.

9.1.4 SAW Chemical Sensor

Figure 9.4 shows a block diagram of a SAW chemical detector. The sample is introduced and chemicals in the sample are sorbed onto the surface. This results in changing the acoustic wave frequency that propagates through the surface. Almost all SAW chemical sensors rely on mass sensitivity of the device. Since the polymer coats the surface of the piezoelectric substrate to adsorb molecules from the vapor, any sorption of chemicals on the polymer will change surface wave propagation on the substrate surface. Wave frequency change and its attenuation are used to

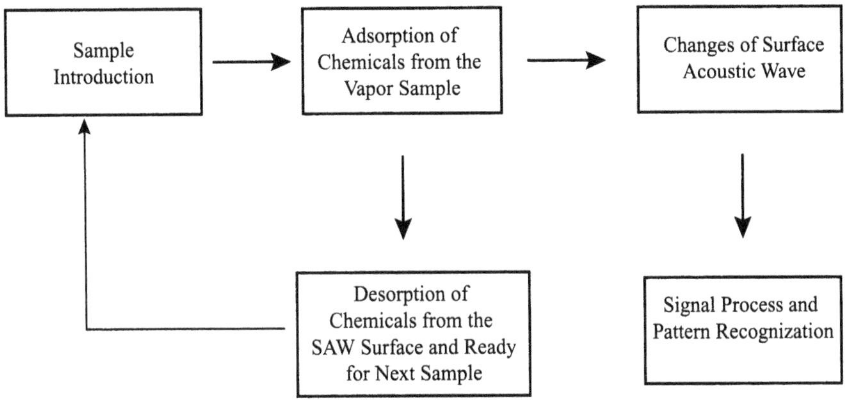

Figure 9.4 Block diagram of SAW analysis procedure.

determine the amount of substance deposited on the device. The signals are processed to identify chemicals sorbed from the sample. Flash heat is then applied to the SAW device to release the sorbed chemicals from the sites. The sensors are ready for the next sample analysis. By assuming that sorption equilibrium is reached, concentration of the chemical in the vapor can be determined.

9.2 INSTRUMENTATION

Chemical detectors using SAW technology are under aggressive development by many different practitioners. As a result, these detectors may be very different from each other. In general, an array of SAW sensors coated with various polymers is used to minimize interference and increase selectivity. A typical SAW detector may have three, four, or more sensors. One of the sensors is isolated from the sample flow path for compensation of temperature-induced effects. The other sensors, coated with specific polymers that are selective to target chemicals, will have different response characteristics when exposed to the sample vapor to detect different types of compounds. A pattern of responses from these sensors can be monitored and an algorithm developed to identify the chemicals sorbed. When the pattern matches the criteria stored in the algorithm, the detector will indicate the presence of the targeted chemical. The array is housed within a package that has a pathway for vapor to flow over all sensors. This package can be temperature controlled to reduce temperature effects and optimize sensitivity.

9.2.1 Sample Input

SAW-based detectors can draw vapor directly or through a sample concentrator before analysis by the sensor array. Sample concentrators are substances with high surface areas to sorb large amounts of incoming molecules from an air sample. Charcoal is commonly used to sorb targeted chemicals from the vapor stream. Charcoal sorbs CWAs and TICs strongly, striping them from the air and, thus, concentrates them on its surface. This concentration technique before the sample is presented to the sensors by heat desorption, is very helpful for achieving better sensitivity and improving the instrument's minimum detectable level (MDL). It not only permits detection of low-concentration samples, but also reduces certain undesirable features of direct-sample detectors. The sample preconcentrator acts as a buffer to reduce effects from moisture and interference chemicals that the SAW sensors are quite sensitive to. The sampling time and rate are controlled so that the concentration can be calculated for detection optimization. Some devices use a gas chromatograph (GC) column in front of the sensors to further reduce interference with targeted chemicals to enhance selectivity. The use of a preconcentrator and GC column increases the sensitivity and selectivity of the detector and further minimizes the abovementioned adverse effects.

The preconcentrator, after sorbing chemicals from the sampled air during the sampling period, is then heated to release the sorbed chemicals. A flow of carrier gas sweeps the released chemicals into the GC column for separation of components prior to reaching the SAW sensor array for detection/identification. Relative elution times for each of component can serve as a means of identification. The relative strength of signals can be related to concentrations of the substance in the sampled air.

9.2.2 SAW Device

A typical SAW device includes two sets of interdigital transducers (IDTs) made by a photolithographic process on the surface of a piezoelectric substrate (Figure 9.5). An electrical field is applied to one of the IDTs to form an acoustic wave on the surface of the sensor. The wave propagates through the substrate to the other IDT where the mechanical energy of the acoustic wave is converted back to electricity.

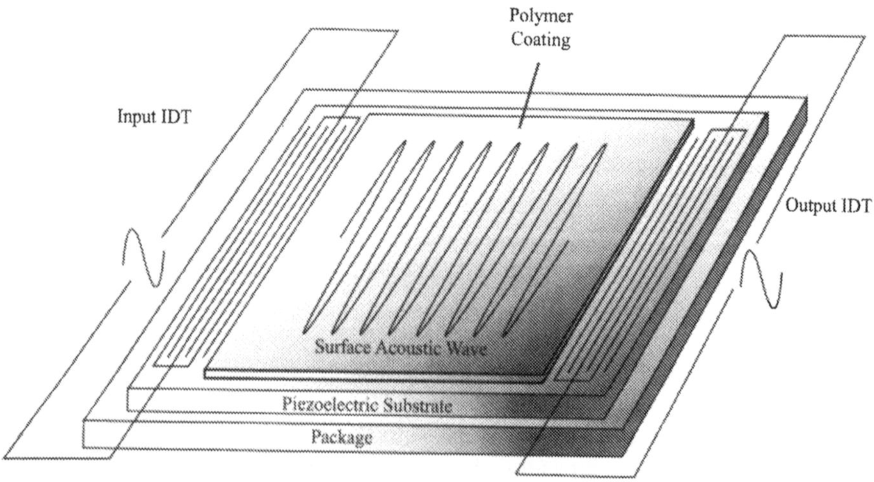

Figure 9.5 Schematic diagram of a SAW device.

To selectively detect targeted chemicals, the propagation path of the acoustic wave is coated with a selected polymer substance. One important requirement for the coating is that the sorption of targeted chemicals must be totally reversible after an analysis. All sorbed substances must be released totally without residue after an analysis so that the detector can be used to analyze the next sample. The sorption coefficient of the coating with respect to the specific type of chemicals should possess high K value to achieve good sensitivity and selectivity. Fast recovery time is also a required characteristic of the polymer.

Practically, it is impossible for a polymer coating on a SAW sensor to be made to only sorb one chemical. A single polymer may sorb several chemicals from a gas mixture. This opens the door for potential false alarms. Similarly, chemicals contained in the gas mixture can be sorbed by more than one polymer sensor in the array with various sorption coefficient values. Each of the sensors in the array is coated with a specific kind of polymer intended for selective sorption of a targeted group of chemicals. The presence of a targeted chemical generates a specific response pattern. Based on the pattern generated, the chemical in the vapor can be identified. Many different sensors having various coatings could be used in a detector to increase its detection reliability.

9.2.3 Temperature Effect and Compensation

The surrounding temperature affects the performance of a SAW device in two ways. Temperature changes the crystalline structure of the piezoelectric substrate that in turn affects the propagation property of the SAW. Temperature changes also affect the sorption and desorption property of the coating film. The effect that affects the property of the piezoelectric crystal can be reduced through installation of a reference SAW sensor that is isolated from the sample vapor. Temperature fluctuation affects not only the SAW sensors for the detection but also the reference sensor. Therefore, the temperature effect on the sensors could be compensated for by negating the similar effect experienced by the reference sensor.

Temperature changes the sorption coefficient of the polymer coating film. The polymer has stronger sorption ability to chemicals at lower temperatures. Consequently, the sensors theoretically should have better detection sensitivity at lower temperatures than those at higher temperatures. However, at low temperatures, much more energy and time are required for effective total release of sorbed chemicals before the next analysis cycle. It is also possible that the sorbed chemicals cannot be totally released at low temperatures, which means that sensors malfunction. In addition, the polymer may become very rigid at extremely cold temperatures and lose its adsorption ability, which also causes the detector to fail.

When temperature rises, the energy of molecules in the vapor becomes higher as the bonding strength between the molecules and polymer becomes weaker. The bonding strength of the polymer may not be strong enough to retain the sorbed molecules striking its surface. This reduces the ability of the sensor to sorb effectively, resulting in a higher minimum detectable level (MDL). This characteristic proves to be a necessary and desirable trait, as it provides a means for fast clearing of sorbed chemicals from the polymer surface after each analysis. The sensors are heated, together with a flow to clear the surface.

For both stationary and handheld detectors, it is possible and desirable to maintain a constant temperature on the sensors during the analysis. Heating the sensor cavity to a temperature above the ambient temperature reduces the effect of ambient temperature fluctuations that are known to influence detection characteristics. Unfortunately, higher-than-ambient temperatures will tend to worsen the sensors' LOD. Therefore, a compromise setting must be chosen to provide the best detection results.

9.3 ANALYSIS PROCESSES

An internal pump draws the sample into the sample preconcentrator where the targeted chemicals are sorbed. After the sampling period — a few seconds to several minutes — the preconcentrator is heated to release the sorbed chemicals. In the absence of a GC column, the carrier gas then sweeps the released chemicals into the sensor cavity where the vapor is evenly distributed onto each of the SAW sensors in the array. The concentration released from the preconcentrator as a slug is much higher than the concentration in the air sample. Chemicals in the released vapor are selectively sorbed onto the SAW sensors in the array and detection occurs. With the help of calibration curves, quantitative or semiquantitative results can be obtained. After the analysis, heat is again applied to the SAW sensors to clear sorbed chemicals. The analysis cycle is repeated. Choosing the proper timing sequences, each of the heating and cooling processes will occur in tandem.

9.4 TECHNICAL SPECIFICATIONS

9.4.1 Selectivity

Because polymers cannot be made specific to only sorb a single chemical, false positive alarms may occur when various unknown chemicals are being sorbed onto polymer. To minimize this false alarm problem, multiple SAW sensors with different polymer coatings are used in a single detector. The combined response patterns can improve detection reliability. For each chemical of interest, the response pattern is established via comparison with standardized samples. The pattern recognition is based on comparison between the output signals and the established response pattern stored in the detector's databases.

In field applications, many unknown components are common. It is very possible that the targeted chemicals found in the field may be mixed with various substances that cause them to behave differently from standard samples used in the laboratory to construct the database. Therefore, false alarms, both positive and negative, are still possible. Hopefully, continued experimentation made by researchers and manufacturers can and will overcome some of these problems.

9.4.2 Sensitivity, MDL, and Response Dynamic Range

SAW detectors can be very sensitive. Sensors may detect chemicals in vapor at the parts-per-billion level. Controlling factors affecting sensitivity are SAW device frequency, polymer properties, and respective targeted chemicals. SAW device using higher frequencies are more sensitive to the change of mass on the polymer coating. As discussed above, polymers with higher adsorption coefficients can sense lower concentrations of chemicals in vapor. It is noteworthy that moisture level in a sample may drastically affect sensor sensitivity. Polymers can sorb water easily. Because water concentration in an air sample can be much higher than the concentration of targeted chemicals, water molecules may occupy many of the active sorption sites of the polymer and effectively reduce sorption ability of the polymer to adsorb targeted chemical molecules. This will significantly reduce detector sensitivity. A preconcentrator, water trap, or GC column helps to minimize or remove water content before the chemical sample reaches the SAW sensors.

The response dynamic range of the typical SAW sensor is very broad and could be up to six orders of magnitude. The sensor reaches its saturation point only when the target chemical in the sample is sufficiently concentrated to occupy most or all active sorption sites available on the polymer-coated surface.

9.4.3 Response Time

Once the chemical vapor comes into contact with the SAW sensors, a response is developed within seconds. A complete analysis cycle for a SAW-based detector includes sampling time, desorption time from the preconcentrator, analysis time by the sensors, and recovery time to clear the sorbed chemicals from the polymer. Additional elution time is added when a GC column is incorporated. Analysis time is only a very small portion of the entire cycle. Detection speed depends on sampling time, especially when a preconcentrator or GC column is used for low-concentration samples. While it is possible to bypass the preconcentrator to reduce the analysis cycle time, the probability of false responses by the sensors significantly increases. Recovery time depends on the quantity of chemicals sorbed, strength of the bond between the sorbed molecules and sorption polymer, and the temperature necessary for releasing sorbed chemicals. Recovery time ranges from a few seconds to several minutes.

In summary, SAW-based chemical sensors have the potential to provide fast and sensitive detection of chemicals. Sensors can be miniaturized and made economically. Therefore, SAW-based detectors can be made into small, lightweight packages. For low-concentration vapor, a preconcentrator alone or a preconcentrator connected in tandem with a GC column is used to increase sensitivity and improve selectivity. Selectivity also depends on the polymer coating on the surface of the piezoelectric substrate together with using an array of several sensors. The current Joint Chemical Agent Detector (JCAD), under development by the U.S. military, is a SAW-based detector. The JCAD intends to replace the current Automatic Chemical Agent Detection Alarm (ACADA), which is a device based on ion mobility spectroscopy.

9.5 APPLICATIONS

A few devices using the SAW technology are described below. The SAW MiniCAD® (Figure 9.6) is a commercially available, handheld, pocket-sized detector that can monitor for trace levels of sulfur mustard and G-type nerve agents with specificity. The instrument is equipped with a sampling pump and a sample concentrator that can be heat desorbed to provide an enriched sample concentration for the two microsensors to analyze. More details are included in the fact sheet on this device at the end of the chapter.

The HAZMATCAD® (Figure 9.7) was developed subsequent to the SAW MiniCAD. The HAZMATCAD measures 2.3 in. × 2.5 in. × 7.8 in. and weighs approximately 1.5 lb. It detects sulfur mustard and G-type nerve agents using micro-SAW sensors, as well as TICs or blood agents (AC, CG, and CK) by including an EC cell for this purpose. This device's operational temperature ranges from −10°C to 50°C at all levels of humidity. The time required for the device to warm up (to a specified temperature) before detection is approximately 30 sec. A fact sheet on this device at the end of the chapter provides more details.

The HAZMATCAD has two modes of detection. The "Fast" mode requires 20 sec per sample cycle. The "High Sens" mode requires 120 sec per cycle of sample and purging. This mode can be used to monitor low-concentration situations. During

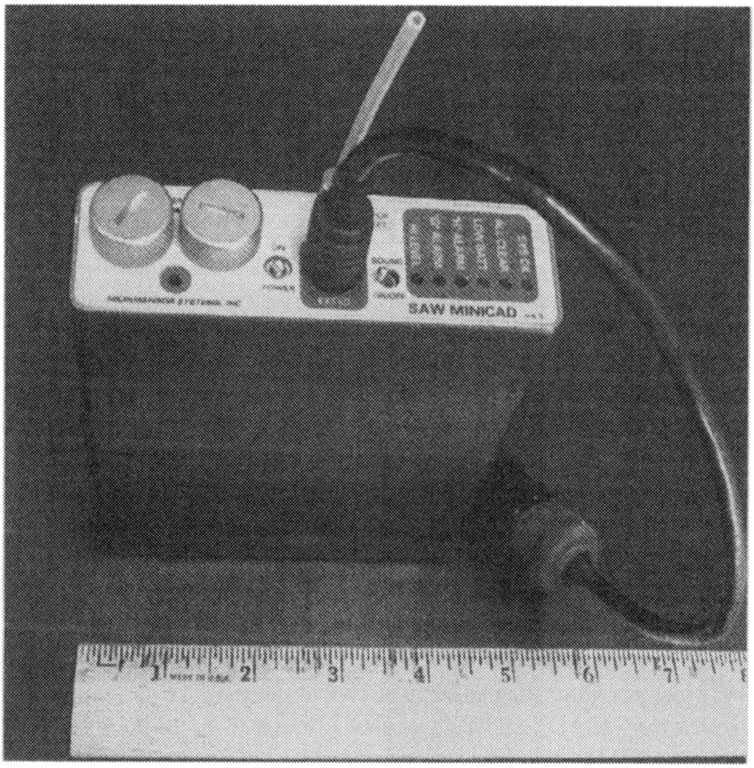

Figure 9.6 MiniCAD detector. Photo courtesy of K. Y. Ong.

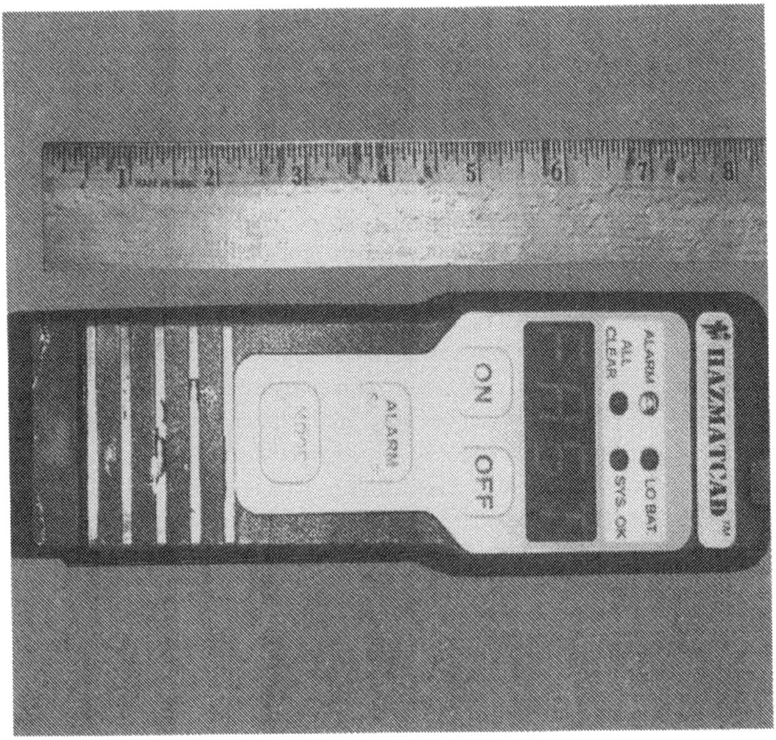

Figure 9.7 HAZMATCAD detector. Photo courtesy of K. Y. Ong.

operation in the high-sensitivity mode, the detector draws its sample for a longer period so that more of the targeted compound is concentrated for the sensors to detect. The instrument does not use a GC column to aid in separating sample components. The use of a heat-desorbed sample concentrator permits the instrument to operate at various humidity levels without significant adverse effects.

Testing conducted under the Domestic Preparedness Program has confirmed that the SAW sensor can detect and differentiate the HD and the G nerve agents. Results indicated that sensitivity toward the more volatile GB is lower than that of the less volatile GA. Humidity effects on detection were avoided with the use of a preconcentrator. Temperature affected some of the detection algorithms due to the insufficient size of the heater used.

The HAZMATCAD technology has further been incorporated in the development of the CW Sentry® system that is used for monitoring buildings or subway systems where the unit can operate continuously around the clock 7 days a week. The SAW sensors fit well in this application because it requires very minimal maintenance and practically no consumables except for an occasional dust filter change.

Another detector, still under development for the U.S. military by BAE Corporation, is the JCAD, which uses SAW techniques to detect nerve, blister, and blood agents, and assorted TIC compounds. When fully developed, it should have the ability to report on concentrations of respective agents. The detector will be lightweight, and handheld or carried in a small pouch. It will be powered by rechargeable batteries that provide more than 12 hr of continuous operation. The JCAD is expected to be operational from −32°C to 49°C at all humidity levels.

The Micro-Chem Lab® detector, developed by Sandia National Laboratories, incorporates a micro-GC column after the sample concentrator to separate sample components before reaching the SAW sensor array for detection and identification. While the elution components are sensed with generic types of SAW sensors, the GC's elusion time required for a detected component could provide identification. This system is also lightweight. As of this writing, this detector was undergoing numerous tests.

Different versions of SAW-based detectors are continuously being developed, as SAW devices can be manufactured at relatively low cost as compared to other technologies. The polymer coating can be refined to provide greater selectivity and/or higher sensitivity. The sensors can be easily miniaturized, and more sensitive models are emerging. Methods to harvest response signals are becoming more mature and sophisticated through aggressive research efforts. It is hoped that a highly reliable CWA and TIC detector based on SAW technology that fits JCAD requirements will be available soon.

9.6 ELECTROCHEMICAL SENSORS

Both HAZMATCAD and JCAD employ EC sensors for detecting blood agents and other TICs. Sensors based on EC techniques or EC sensors have been used to measure toxic gases at the parts-per-million level for many years. This technique is based on the electric current generated by the oxidation or reduction reactions. Electrochemical sensors have properties that make them attractive for long-term monitoring, real-time measurement, process control, and safety applications. Many different gases, such as CO, O_2, and H_2S, can be detected and monitored in real time with EC sensors. In fact, any gas that undergoes oxidation–reduction reactions can be detected with EC sensors. Since sensor structure is very simple, it can be made very small and requires little power for personal monitoring applications.

9.6.1 Oxidation–Reduction Reactions

EC sensors utilize the electric current generated by oxidation or reduction reactions between chemicals in the air and the electrolyte residing in the sensor. The reaction involves the exchange of electrons. As an example, a simple oxidation–reduction reaction is illustrated through the burning of hydrogen gas in air. When hydrogen burns, it reacts with oxygen in the air to form water molecules.

$$2H_2 + O_2 \rightarrow 2H_2O \tag{9.6}$$

This reaction can be rewritten in its ionic format that shows the exchange of electrons between the hydrogen and oxygen molecules:

$$2H_2^0 + O_2^0 \rightarrow 2H_2^{+1}O^{-2} \tag{9.7}$$

Each hydrogen atom donates its single electron and forms a positively charged ion,

$$H^0 \rightarrow H^{+1} + e^{-1} \tag{9.8}$$

while each oxygen atom gains two electrons and forms a negatively charged ion:

$$O^0 + 2e^{-1} \rightarrow O^{-2} \tag{9.10}$$

The lost electron from the hydrogen atom flows to the oxygen atom. Thus, the oxygen in this reaction is the "oxidizing agent" or "oxidant" that oxidizes the hydrogen. The hydrogen loses its electron, and thus is a "reducing agent" or "reductant." In this oxidation–reduction reaction process, the "oxidizing agent" (oxygen) is reduced to a lower valence by the "reducing agent" (hydrogen). Oxidation and reduction always occur simultaneously. A common application relying on oxidation–reduction reactions is to generate electric power in a car battery.

An EC sensor is similar to an electric generator. The oxidation or reduction of the electroactive gases with the residing reagent generates a measurable electric current in the sensor. This generated electric current, instead of being used to produce power, is measured as a detection signal.

9.6.2 The Sensor

Figure 9.8 shows the structure of a typical EC sensor that consists of three electrodes. Although EC sensor types are numerous, their basic structures and principles are similar. Basically, three electrodes — working electrode (WE), counter electrode (CE), and reference electrode (RE) — are immersed in an electrolyte, such as H_2SO_4 solution in car batteries. Solid-state electrolytes are also used. The electrodes and the electrolyte are included in a small (e.g., plastic) container. The WE, also called the sensing electrode, consists of a porous polymer membrane that coats a small gold or platinum plate, which in turn is connected to an external electric source at a certain electric potential. The oxidation or reduction of the targeted chemicals occurs when the chemicals are in contact with the electrode. The reaction that occurs here is only half of the oxidation–reduction reaction. The other half occurs on the CE that is used to balance the reaction on the WE. The RE is used to control the electric potential of the WE through a feedback mechanism.

When the sensor is exposed to the sample vapor, molecules diffuse through the porous membrane and come into contact with the WE. Both the membrane and electrode are immersed in the electrolyte. The porous membrane is saturated with the electrolyte. The reaction takes place on the surface of the working electrode where the incoming molecules become oxidized or reduced by the electrolyte. The gold or platinum plate itself serves as the catalyst for the reaction. The reaction generates an electric current that is measured for detection purposes.

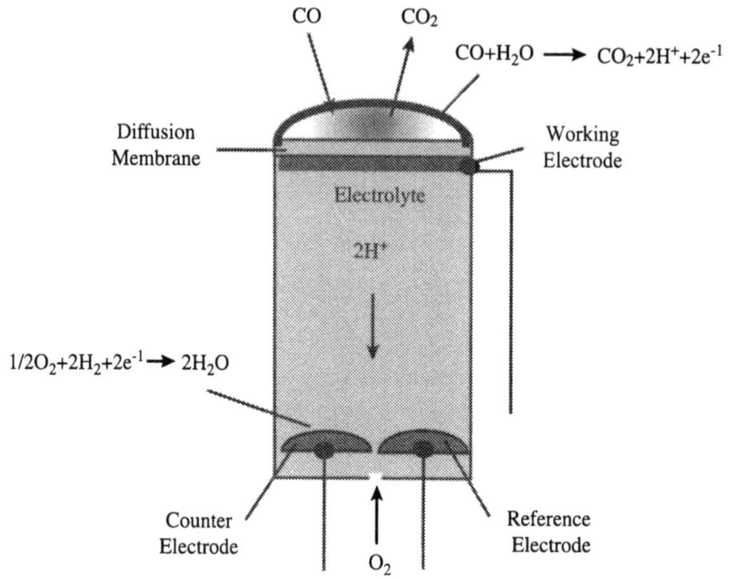

Figure 9.8 Schematic diagram of EC sensor.

As an illustration, we will use the carbon monoxide oxidation and reduction reaction. When CO is in contact with the WE, the CO molecules will be oxidized:

$$CO + H_2O \rightarrow CO_2 + 2H^+ + 2e^{-1} \tag{9.11}$$

The oxidation of CO is half of the reaction. The produced CO_2 is released back to the air while H^+ stays in the electrolyte and migrates toward the CE where the other half of the oxidation–reduction occurs as follows:

$$\frac{1}{2}O_2 + 2H^+ + 2e^{-1} \rightarrow H_2O \tag{9.12}$$

The function of the CE is to complete the oxidation–reduction reaction. Thus, the overall EC reaction is

$$CO + \frac{1}{2}O_2 \rightarrow CO_2 \tag{9.13}$$

The electric current generated by this oxidation–reduction reaction is measured for determining the CO concentration in the sample.

To accurately measure a chemical concentration in the air, the relative voltage between the WE and CE should remain constant. The H^+ ion produced from the WE dissolved in the electrolyte changes the electrical potential of the electrolyte and, therefore, the voltage of the WE relative to the CE. To compensate for this voltage fluctuation due to the change of ion concentration in the electrolyte, a third (reference) electrode is used. The RE is made with material that is non-conductive and stable in the electrolyte. It measures the electric potential of the electrolyte during the analysis and feeds the information back to the WE so that the relative potential difference is adjusted accordingly.

When an EC detector is used for monitoring or detection, the sample flows over the sensor and contacts the surface. The molecules in the vapor sample diffuse through the porous polymer on top of the gold or platinum plate where the oxidation–reduction occurs. This diffusion through the polymer is the controlling factor for detection, and serves to prevent the sensor from becoming saturated by the targeted chemicals. This will ensure that the response is proportional to the concentration of the target chemicals in the sample. The reaction is catalyzed by the gold or platinum surface of the plate. The generated electric current is sent to the processor for signal analysis.

9.6.3 Operation, Sensitivity, and Selectivity

Electrochemical detection of CWAs can be illustrated with an example involving the reaction of a cholinesterase with nerve agents. A solution of cholinesterase enzyme is contained within the cell. When a nerve agent comes into contact with the enzyme solution as it diffuses through the membrane, a certain amount of the enzyme will be inhibited while the uninhibited enzyme reacts with another reagent to form an electrochemically active product that can be measured. Given that the initial amount of enzyme is known, the amount of inhibited enzyme as derived from measurement of the uninhibited enzyme can be related to the concentration of the detected nerve agent.

Other EC sensors measure the change in resistance upon absorbing targeted chemical agents into the thin-film polymer coating on a gold substrate. The swelling or contraction of the film as the agent is absorbed into or released from the polymer causes differentials in resistance measurements for a given applied potential.

EC sensors are relatively sensitive, as they react to chemical vapor concentrations at the low parts-per-million level. However, EC sensors are not as selective as colorimetric detectors (see Chapter 10). They may respond to various chemicals simultaneously without differentiation capability. This is because the oxidation–reduction reaction between the chemicals in the sample and the electrolyte controls the detection. Any chemicals contained in the sample that will react with the electrolyte on the working electrode surface will generate electrical current and are detected together with those from targeted chemicals. Using a chemical filter may reduce or eliminate some of the chemical interference potential. The use of a second working electrode that responds to different sets of chemicals from the first working electrode within the same sensor may also lead to better selectivity.

Several factors may affect the performance of EC sensors. Environmental conditions are major concerns. During operation in high-humidity conditions, the cell electrolyte may absorb water from the air. In contrast, while operating in low-humidity conditions, the cell electrolyte may lose water. This gain and loss of water will change the concentration and volume of the electrolyte within the sensor and affect sensor sensitivity.

Sensor lifetime can be reasonably long. However, its sensitivity may vary due to loss of electrolyte and contamination of the catalytic surface.

The first chemical agent detector that was developed by the U.S. military, the M43, is an EC detector that senses the reaction of a nerve agent with an oxime solution. The electrical signal generated is measured through the use of platinum and silver electrodes.

Another example of chemical agent detection using an EC cell is the Individual Chemical Agent Detector®. The ICAD can detect both nerve and mustard agents simultaneously through the use of two sensor cells.

9.7 FACT SHEETS FOR SELECTED SAW AND/OR ELECTROCHEMICAL DETECTORS

Detector Name

HAZMATCAD and HAZMATCAD/Plus

Manufacturer/Distributor and Contact Information

Microsensor Systems
62 Corporate Court
Bowling Green, KY 42103
Tel: 270-745-0099
Fax: 270-745-0095

Technique Description

This instrument is a handheld SAW-based detector that uses the change in SAW frequency caused by adsorption of substances on the surface of polymeric sensors. The detector has two sensitivity modes: fast mode has a detection cycle of 20 sec per analysis; and high-sensitivity mode requires 1 min per analysis. The detector automatically recycles after each analysis cycle. It can detect nerve, blister, and blood agents. TICs and blood agents are detected via EC sensors.

Chemical Detection Capability and Performance

The detector is designed to detect nerve, blister, choking, and blood agents, together with a list of TICs. Sensitivity is expected to approximate current JSOR requirements. It includes a sample preconcentrator to increase sensitivity and stability.

Other Features

Interface for network connection. The HAZMATCAD/Plus uses four EC sensors to detect blood and choking agents as well as some TICs.

Domestic Preparedness Test Report

Domestic Preparedness Program Testing of HAZMATCAD Detectors Against Chemical Warfare Agents Summary Report, February 2002 (http://hld.sbccom.army.mil/ip/reports.htm#detectors)

Detector Name

Individual Chemical Agent Detector (ICAD®)

Manufacturer/Distributor and Contact Information

Smiths Detection
2202 Lakeside Blvd.
Edgewood, MD 21040
Tel: 410-510-9100

Technique Description

This instrument is a miniature chemical agent detector, approximately the size of a cigarette pack weighing about 8 oz, that can detect and alarm to nerve, blister, and blood agents. It may be used as a point detector that is worn by individuals, and may be connected to a network. The chemical agent concentrations in the electrolyte are measured by multiple-electrode EC sensor systems that trigger an alarm when a preset threshold level is reached.

Chemical Detection Capability and Performance

The detector has two EC sensors covered by a thin diffusion membrane. One sensor detects GA, GB, GD, AC, and CG, and the other detects H and L agents.

Other Features

Visual and auditory alarms. Interface for network connection. Operational life of approximately 4 months depending on temperature during use. EC cells can be replaced in the field.

Detector Name

JCAD Joint Chemical Agent Detector

Manufacturer/Distributor and Contact Information

BAE Systems
6500 Tracor Lane
Austin, TX 78725
Tel: 512-929-4753

Technique Description

Instrument is a handheld detector using change in SAW frequency caused by adsorption of substances on surface of polymeric sensors. Electrochemical sensors are used to detect TICs.

Chemical Detection Capability and Performance

Instrument is designed to detect nerve, blister, and blood agents together with a list of TICs. Sensitivity is expected to meet current JSOR requirements. It will include sample preconcentrator to increase sensitivity to parts-per-billion level.

Other Features

Interface for network connection. This device is being developed by the U.S. Joint Services as a future detector to replace/complement the ACADA.

Detector Name

MiniCAD SAW MK

Manufacturer/Distributor and Contact Information

Microsensor Systems
62 Corporate Court
Bowling Green, KY 42103
Tel: 270-745-0099
Fax: 270-745-0095

Technique Description

This instrument is a pocket-sized detector that uses the change in SAW frequency caused by adsorption of substances on the surface of polymeric sensors. The detector has two sensitivity modes: fast mode has a detection cycle of 20 sec per analysis; and high-sensitivity mode requires 1 min per analysis. The detector automatically recycles after each analysis. It can detect nerve, blister, and blood agents. Blood agent detection is done via EC sensors.

Chemical Detection Capability and Performance

This detector is designed to detect nerve and blister agents. Sensitivity is expected to be around the IDLH region. Sample preconcentrator is included to increase sensitivity and stability.

Domestic Preparedness Test Report

Domestic Preparedness Program: Testing of SAW MiniCAD MKI Detector Against Chemical Warfare Agents Summary Report, September 2001 (http://hld.sbccom.army.mil/ip/reports.htm#detectors).

CHAPTER 10

Colorimetric Technology

Colorimetric detectors analyze the color change resulting from the reaction between the targeted chemical and reagents. This technique has been used in many different applications. Good examples of colorimetric detection are the use of litmus paper (or pH paper) to test the pH value of a solution, and the common water test kit used to analyze chlorine concentration in swimming pools. The litmus paper test is an example of a direct reading: the color is formed by the reaction with reagents impregnated on the paper, and no further manipulation occurs. As soon as the litmus paper is wetted with the solution, the pH of that solution causes the impregnated reagents to change color according to the solution's pH value. Chlorine content testing of swimming pool water is based on the addition of reagent o-Tolidine to the solution to form the visible orange color that can be used for quantitative determination. Color intensity is related to the relative concentration of chlorine in the water. Colorimetric detectors using different methods are all easy to use, low cost, and provide relatively fast responses.

Colorimetric detectors often take the form of badges and tubes where the signal (color change) is detected by the human eye instead of electronic devices. Therefore, most colorimetric detectors can be made small and simple, are less costly to produce, and do not use electric power for operation.

Most colorimetric detectors are designed to be selective because of carefully selected reagents. The reagents are selected to react only with a specific class of chemical compounds that produce the proper color change for detection purposes. The number of interferences for the specified detection is usually very limited. Typically, a single sensor is used to detect or monitor one class of chemicals. For example, a badge that is designed for monitoring phosgene will detect the presence of phosgene only and cannot be used for the detection of nerve gases. Thus, the advantage of this detector is its lower false alarm rate. On the other hand, it has the disadvantage of requiring many sensors for field applications.

To overcome the technique's major disadvantage, some detectors have incorporated several sensors for detecting specific classes of compounds. One badge contains several sensors, or several different detection tubes are used for each surveillance-monitoring task. For example, the military M18® detection kits are assembled with various sensor spots and tubes to detect CWAs. The later M256® Chemical Agent Detector Kit, using similar detection methodologies, combines all required reagents for various types of detection into a single sampler to detect CWAs.

Monitoring badges and detection tubes are mainly used for qualitative analysis or semiquantitative analysis, which uses human eyes as the signal (color) processors. The sensitivity of the human eye varies widely. Each person has slightly different color perception, and many people suffer from some degree of color-blindness. Environmental conditions and color formation media also affect colors and their interpretation. It is difficult to observe color in dim or bright light situations. These factors limit the effectiveness usage of detection badges and tubes as a reliable quantitative technique.

Since water can enhance or delay reactions, the moisture content of sensors, badge, tube, or paper-tape should be kept at a certain level. When manufacturing test kits, concerns for proper moisture level must be addressed. During use or storage of the kits, moisture may be lost from or absorbed by the sensor spots, which can affect the detection to become unreliable. Techniques to counter effects are usually incorporated when this methodology is employed.

10.1 PRINCIPLE OF OPERATION

The principle underlying the colorimetric method of detection is quite simple. Detectors are made with sorbent substrates, paper or paper tape, or glass tubes that are impregnated with colorimetric reagent(s), which react with a specific targeted chemical to form a particular color. Colorimetric tubes may consist of finely impregnated beads together with the necessary reagent ampoules. This arrangement enables the sample flow to contact the reagent-impregnated surface more effectively and homogenously.

Figure 10.1 is a block diagram showing the principle of operation. When the targeted chemical in the sample comes in contact with the chemically impregnated surface, whether through a passive or active sampling technique, the surface color changes due to specific chemical reactions. The change in color indicates the presence of the target chemical(s). Concentration of the targeted chemical in the sample can be estimated based on the intensity of the color (monitoring badge and paper tape) or the length of the stain (colorimetric tube) developed over the exposure time. Common types of colorimetric detectors are described below.

10.2 INSTRUMENTATION

The three common colorimetric-based detectors are badges, tubes, and paper tape. They can be very simple without any moving parts, that is, badges or paper

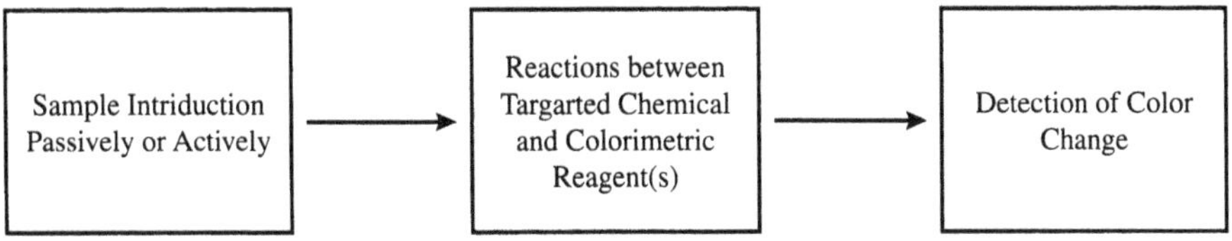

Figure 10.1 Block diagram of colorimetric detector analysis procedure.

spots. Some will use a sampling pump that can be powered by electricity or manually operated to draw sample into the detection tube. More advanced systems involve an automatically operated advancing tape, reagent injections, and colorimeter unit to detect color formation. Each type is designed for specific applications.

10.2.1 Badge Monitoring Kit and Paper Spot Detector

Colorimetric badges and paper spot detectors are based on papers that are impregnated with reagent before exposure to air. Both use a passive sampling technique, that is, the sensors are exposed to the air without any other means to force the air into contact with the sensor. Some sensors may require additional reagent and manipulation after exposure to form the color. This type of monitoring kit is inexpensive, lightweight, and easy to use. They can produce direct color development results without requiring a pump or any moving parts for sampling. Monitoring kits that are made as badges, used generally for personal exposure monitoring, are worn by individuals. They may also be used for area monitoring. Unlike other types of detectors, these kits are consumed after a single use. Monitoring results show only total exposure dose during the exposure time.

During the monitoring, the impregnated badge or paper spot is exposed to the air. Targeted chemicals, if any, diffuse onto the surface of the sensing area and react with the reagents. The reaction between the targeted chemical and the reagent changes the color of the badge/paper, which indicates presence of the target chemical. This type of monitoring kit is usually precalibrated to permit determination of the exposure level. The resultant color is compared to the standard color pallet similar to the function of a calibration curve. The results are, at best, semiquantitative.

10.2.2 Detection Tube

A detection tube can be made by coating the inner wall of a thin transparent tube with reagent. Similarly, fine impregnated beads of adsorbing materials may be packed inside of the tube (Figure 10.2). The turbulence created by the beads in the sample flow path enables a more complete reaction between the chemicals in the vapor and the reagent on the surface of the beads. Another type of colorimetric tube consists of adsorbent beds and reagent ampoules sealed within a glass tube. The reagent ampoules are broken to provide the necessary reactions that produce the color for detection. Although the monitoring badge and spot paper and the detection tube share similar colorimetric principles, there are important differences among them. Sample tubes require a suction pump to draw air samples, while the badge

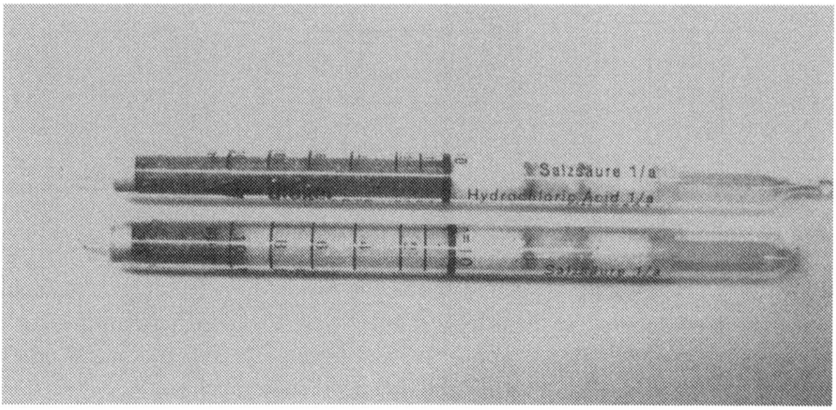

Figure 10.2 Colorimetric tube.

and spot paper is passively exposed to the air. The tube is not designed for continuous monitoring. It is used as a means to detect the presence of targeted chemicals at test time. In general, tubes can perform detection with greater sensitivity in shorter time.

The tubes are sealed until ready for detection to avoid any potential contamination. At the time of detection, both ends of the tube are opened and the tube is connected to the special pump (Figure 10.3). The pump can be a manually operated suction pump or a battery-operated pump that draws air at a certain flow rate. The coated reagents will react with the targeted substances to produce the stain on the sensing surface in direct-reading types of detectors (Figure 10.2). Normally, a colorimetric tube requires a fixed volume of sample for quantitative analysis.

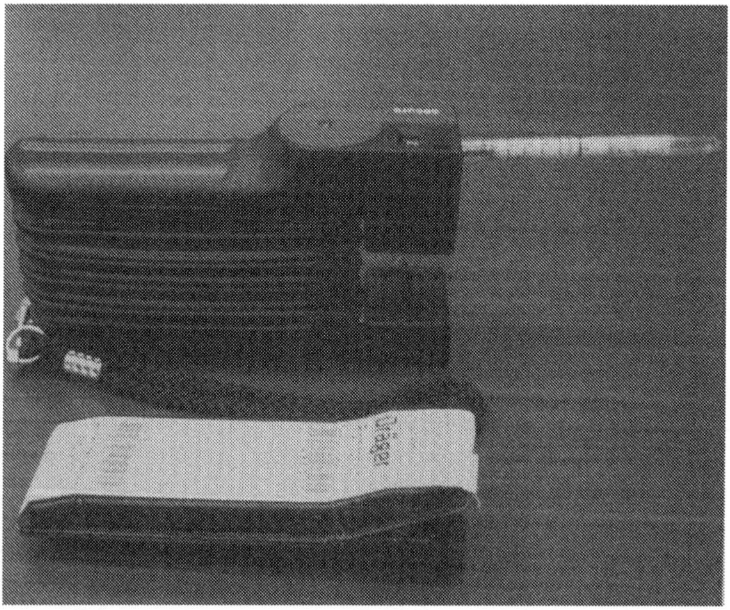

Figure 10.3 Colorimetric tube with sampling pump.

The targeted chemical reacts with the colorimetric reagent first in the area closest to the inlet of the tube; thus, the most intense color is developed at that end. When more of the targeted chemical is drawn in, the reagent near the inlet may become mostly consumed by the target chemical, so more of the target chemical is available to penetrate farther into the bed and produce the stain there. Thus, the length of the colored stain increases. This type of colorimetric tube is usually precalibrated with marks at different positions on the tube to indicate the approximate concentration of the target chemical in the air. The reading from the tube after detection is related to total targeted chemical dosage. Sampling time and flow rate are considered when calculating estimated vapor sample concentration.

Sensitivity of the colorimetric tube varies with the sampling volume and is generally in the parts-per-billion (ppm) to parts-per-million (ppm) level. For less concentrated vapor, a larger sampling volume introduces more of the chemical for the reaction, therefore, sensitivity increases. The dynamic range of a tube is usually limited with respect to the total amount of targeted chemical that it is exposed to. The operating procedure provided by the manufacturer is usually chosen to achieve the intended detection thresholds for which the tubes are designed. Different tube types may use a different number of strokes.

Another type of colorimetric detection tube uses solid adsorbent to adsorb the incoming targeted chemicals either with or without first being impregnated with reagents contained in an ampoule. Another reagent is added after the sample is adsorbed to the dry adsorbent or absorbed to the reagent-wetted adsorbent material to form the color. Depending on the reaction mechanism, appearance or disappearance of a given color is used to denote detection.

10.2.3 Advancing Tape–Based Colorimetric Detector

These colorimetric devices use tape impregnated with chemical reagents as a platform for detection. The tape is commonly about a half-inch wide and made of paper or cloth. Unlike the monitoring badges discussed above that are exposed to the air and take samples passively through diffusion, the advancing tape instruments draw samples through the reagent-impregnated tape using a pump at certain flow rate. After the predetermined sampling time, the tape is advanced to the next stop where a color-developing reagent is added and the color development is monitored by a colorimeter. The rate or intensity of the color developed or the fading of a developed color determines the concentration of the target chemical.

Major components of an advancing tape–based colorimetric detector include colorimetric tape, sample pump, and colorimeter (Figure 10.4). The tape that is or will be impregnated with the reagent serves as a base to generate the color signal. The tape must be isolated from exposure to the sample air except for a small spot that is used as the detection sensor. The tape is advanced at a predetermined time. This allows a new “sensor” spot to be exposed to another sample. The air pump draws the sample vapor through the tape at the spot that has been impregnated with the reagent from the dispenser so that the targeted chemical reacts with the impregnated reagents during the sampling cycle. If the sampled air contains the targeted chemical, reactions between the reagent and the targeted chemical will

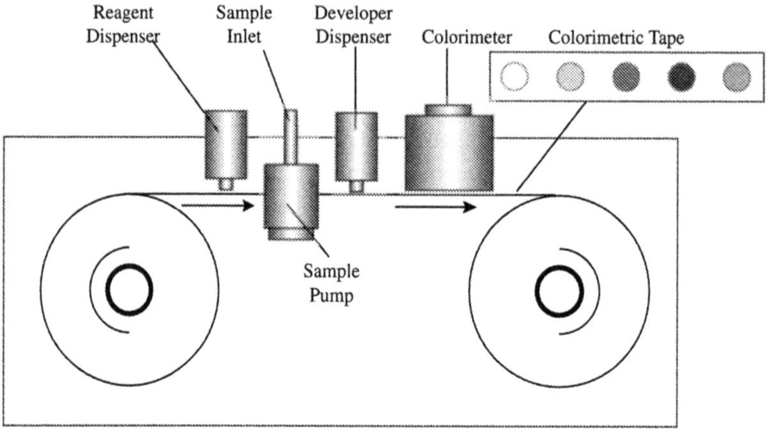

Figure 10.4 Schematic diagram of tape-based colorimetric detector.

change the color of the spot. For some chemicals, the color spot must be developed by adding another reagent. The spot will then be monitored by the colorimeter to record the color observation. This results in one sample observation per tape advancement period thereafter. The sample cycle is dependent on the time required for each step to ensure sufficient time for reaction and color development. Therefore, the cycle time would be the longest time required by one of the steps.

An electronic device called a colorimeter that measures the color intensity is used to determine the intensity of the color that can be related to concentration. This colorimeter measures the light reflected from the exposed stain as well as the light reflected from an unstained area as the reference. The difference between the two signals is used to determine detection. Alternatively, the rate of color formation or color fading could also be used. The instrument can first measure the color at the time the developer reagent is added as compared to the subsequent color intensity after a time interval. This type of instrument usually includes temperature controls that accelerate the rate of reaction.

Another method that monitors the color change rate instead of color intensity at the end of each sampling period is also used. An intensity threshold is preset, and stain intensity is continually monitored. An alarm is triggered as soon as the stain reaches its preset threshold. This rate of change is relative to concentration of the targeted gas in the sample; thus, it can be used to estimate sample concentration. At higher concentrations, the stain intensity reaches the defined threshold in a shorter time than at lower concentrations (Figure 10.5). The advantage of this method is that when high concentrations of the targeted substance are detected, the device quickly delivers a response, thereby speeding up the alarm time.

This colorimetric tape method provides much faster detection of low-concentration samples than badges or tubes could provide. Its LOD can be as low as parts per billion. More importantly, this method provides continuous detection capability that tubes or badges lack. It also records exposure history.

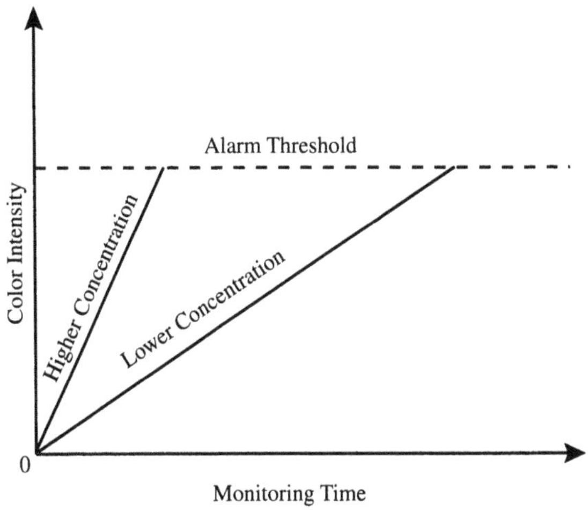

Figure 10.5 Detection based on change rate of color intensity.

In conclusion, the colorimetric detection technique is, in general, very selective and sensitive. Three different types of colorimetric detectors — badges, tubes and papers, and moving tape — use a similar chemical reaction process but with different procedures. Badges are typically used to monitor total exposure levels (dose monitoring), called dosimeters. Tubes or paper spots are used as point sampling detectors that could provide approximate concentrations. Moving tapes are used in detectors designed for continuous monitoring. By combining several different tubes or paper spots, a colorimetric detector can detect multiple chemicals.

10.3 APPLICATIONS

The most well-known colorimetric detectors for CWA detection are the M8® Chemical Agent Detector paper (Figure 10.6), M9® Chemical Agent Detector paper, and the M256A1® Chemical Agent Detector Kit. The M8 and M9 papers each contain a dye that produces a color when dissolved via contact with liquid chemical agent droplets. The M9 paper shows generic red spots, while the M8 paper distinguishes among three classes of agents. M8 paper is used to test suspect droplets of liquid on a surface. The paper will turn yellow when in contact with a G-type nerve agent, red when in contact with mustard agents, and dark green when in contact with a V-type agent. The use of M9 paper is limited to showing red dots in contact with an agent aerosol, warning of potential exposure to a CWA attack. Unfortunately, such M9 detection per se would not protect the individual from harm by the target agent. Both M8 and M9 CWA detector papers are examples of direct colorimetric detection.

The M256A1 Chemical Agent Detector Kit (Figure 10.7) includes M8 paper and a vapor sampler that uses colorimetric detection technology in different ways. It consists of paper spots that are impregnated with reagents (or the spots are impregnated prior to exposure to the air) for simultaneously detecting the presence

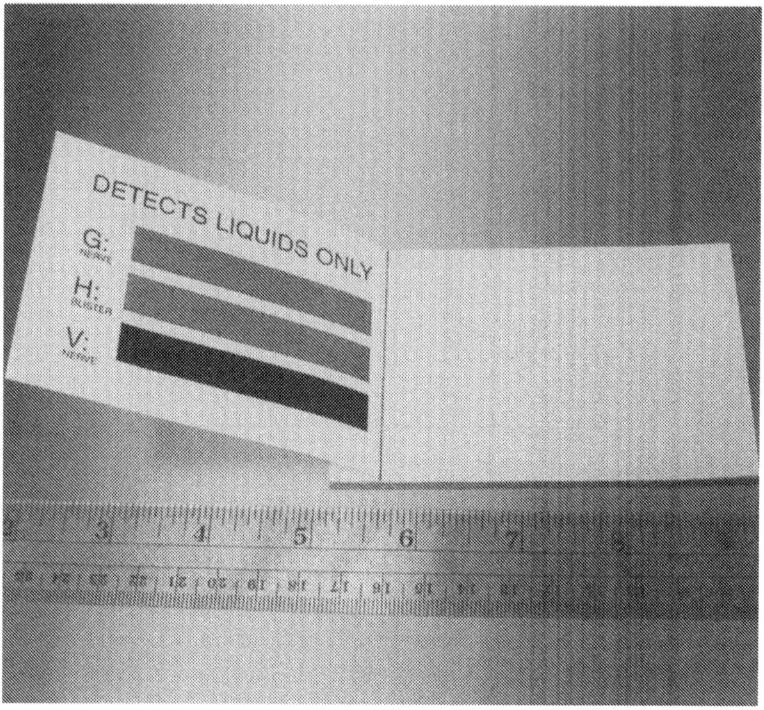

Figure 10.6 M8 paper detector.

of multiple CWAs. All required reagents are encapsulated in sealed ampoules arranged in order according to reaction requirements. The ticket can detect nerve, blood, and blister agents from a single exposure. Two components produce a direct color development without further manipulation after the exposure. The other two spots require the addition of color development reagents to produce colors after exposure.

The M256A1 Chemical Agent Detector Kit is generally used for postincident monitoring to determine the level of CWA presence (and thus when it is no longer a threat) because of its ability to detect very low concentrations of nerve agents. Nerve agents are detected using an enzyme reaction with the anticholinesterase effect that is characteristic of these agents. The M256A1 Kit is also used to verify the presence of CWAs when one or more other generic CWA detectors indicate their potential presence. Because of the lengthy time period required for detection using the M256A1 Chemical Agent Detector Kit, it is not a suitable device for early warning of CWA exposure.

The well-known M18A1® Kit, Draeger tubes (Figure 10.2 and Figure 10.3), and detection tubes from manufacturers such as the Mines Safety Appliances (MSA), among others, consist of sealed glass tubes that contain the required reagents and support beds to test for the presence of targeted classes of substances. Numerous detector tubes are available, covering the range of most TICs as well as CWAs.

Figure 10.7 M256A1 Kit. Photo courtesy of K. Y. Ong.

The Domestic Preparedness Program has tested two types of these tubes for CWA detection ability. For example, the thioether tubes can detect HD, and phosphoric acid ester tubes can detect nerve agents at very low concentrations.

The use of moving-tape instruments to develop the color that can be monitored with an electronic colorimeter can be best illustrated by the Russian enzymatic detector used by the Aum Shinrikyo group that produced the sarin gas in the Tokyo subway incident in 1995. That instrument is an automatic nerve agent detector; acetylcholine or acetylthiocholine is hydrolyzed and then reacts with a color developer to form the color in the absence of a nerve agent. The instrument utilizes a moving tape methodology in which the sampling spot is first impregnated with an enzyme solution before the sample air is passed through the spot. After a sampling period, the cloth tape is advanced to move the exposed spot to the next stop where the reagent containing the color developer that reacts with the hydrolyzed enzyme and acetylcholine or acetylthiocholine product to form the color. In the presence of a nerve agent, the anticholinesterase effect of the nerve agent inhibits the cholinesterase enzyme such that the reaction with the acetylcholine or acetylthiocholine is deterred. Consequently, less color results.

The U.S. military has also used similar enzyme detection techniques, but does not depend on moving tape for CWA detection, which was known as the real-time monitor in the 1970s. This is another example of how colorimetric detectors can be used in CWA detection. The system was developed for safety monitoring of workplaces where nerve agents were being disposed of to support the CWA demilitarization program. The system uses a moving solution system. An air sample is drawn through a scrubber concentrator that is wetted with a pH-adjusted aqueous solution. A portion of the sampled solution is pumped into the tubing via the moving solution system where it encounters an injection of cholinesterase enzyme solution. The presence of nerve gas will inhibit the enzyme as it moves along the tubing; color intensity varies according to nerve agent concentration. By comparing the baseline color (without nerve agent) and the resultant color (after exposure to nerve agent), the relative concentration can be calculated, unless the concentration is so high that the enzyme was totally inhibited.

Colorimetric detectors are usually specifically developed for certain groups of compounds that undergo similar chemical reactions. Therefore, they are usually less prone to interference compared to other types of generic detection technologies. The downside is that reactions require time and manipulation, and different reactions are required to detect different types of compounds. Therefore, colorimetric detectors are best used to supplement other types of devices that could provide more effective real-time detection of CWA or TIC presence.

10.4 FACT SHEETS ON SELECTED COLORIMETRIC TECHNIQUE–BASED DETECTORS

Detector Name

Chemical Agent Liquid Detectors, 3 Way, and M8 and M9 Paper

Manufacturer/Distributor and Contact Information

Tradeways Ltd.
184 Duke of Gloucester Street
Annapolis, MD 21401
Tel: 410-295-0813

Technique Description

This M8 is designed to detect the presence of chemical agent liquid droplets via colorimetric reactions. Provides rapid identification of the three major groups of CWAs, V, G, or H. M9 contains dye that is dissolved when hit with agent aerosol to show red dots as an indication of an exposure to aerosol.

Chemical Detection Capability and Performance

The M8 paper turns green upon exposure to V-agents, yellow when G-agents are present, and red when exposed to H. The M9 paper does not differentiate among the three agent types. Liquid agent droplets dissolve the impregnated dye into red dots for detection purposes.

Other Feature

Not suitable for CWA vapor detection. M9 paper is worn by soldiers intended to detect an aerosol exposure only. M8 is used as a wipe to identify liquid droplets as V, G, or H agent-class.

Detector Name

Chemical Agent Water Test Kit: M272®

Manufacturer/Distributor and Contact Information

Tradeways Ltd.
184 Duke of Gloucester Street
Annapolis, MD 21401
Tel: 410-295-0813

Technique Description

The Chemical Agent Water Test Kit M272 consists of colorimetric detector tubes and enzyme impregnated detection tickets for nerve agent detection. The kit is designed to detect the presence of chemical agents in water samples.

Chemical Detection Capability and Performance

The detector sets are designed to serve as postincident management monitoring of CWAs or TICs. Sensitivity is expected to meet current drinking standard requirements. Processing time varies according to the number of manipulations and the required reaction time for color development from each sample. The time required for the process to complete is quite lengthy. Sensitivity is limited by the ability of the Alka Seltzer®–type tablets to drive agent vapor out of the solution.

Other Feature

Visual indication of detection through the color development by human vision.

Detector Name

M18A3 and M256A1 Chemical Agent Detector Kits

Manufacturer/Distributor and Contact Information

Tradeways Ltd.
184 Duke of Gloucester Street
Annapolis, MD 21401
Tel: 410-295-0813

Technique Description

These kits detect CWA vapors through the use of colorimetric reactions on reactant-impregnated paper or in detection tubes. They are designed to detect the presence of chemical agents in the air through the use of colorimetric reactions. Similar techniques are employed in both kits for detection of nerve, blister, blood, and choking agents. The M256A1 Kit permits simultaneous multiagent detection, whereas the M18A3 Kit requires separate use of tubes or paper tickets for each class of agent.

Chemical Detection Capability and Performance

Papers are designed to rapidly detect the existence of CWA agents in liquid. The kits are sometimes used to determine when the concentration of a given CWA is low enough to permit removal of protective facemasks after an incident or during a decontamination operation.

Other Features

Not suitable for gross-level CWA vapor detection due to time required for determination, as the process requires approximately 15 min to complete.

Detector Name

Simultaneous Test Sets from Draeger Safety

Manufacturer/Distributor and Contact Information

AFC International, Inc.
715 SW Almond Street, Suite C
DeMotte, IN 46310
Tel: 800-952-3293

Technique Description

These products are available in several different sets of colorimetric detector tubes. Each set is designed for detection of specific TICs, TIMs, or CWAs at or below IDLH or close to AEL levels.

Chemical Detection Capability and Performance

The detector sets are designed to serve as postincident management monitoring of CWAs or TICs. Sensitivity is expected to meet/exceed current IDLH requirements. Several tubes can be used to simultaneously draw samples. Processing time varies according to number of manipulations and required reaction time for color development. These tubes are not suitable for gross-level detection due to the lengthy time required for the process to complete. Sensitivity can be increased through longer sampling times or increasing the number of pump strokes.

Other Features

Multicompound detection capability varies according to the number of tubes used. Kits are available to meet Homeland Defense needs for TIC or CWA detection.

CHAPTER 11

Photoionization and Flame Ionization Detection Techniques

Photoionization and flame ionization are common detection techniques used for gas chromatographic (GC) systems in laboratory environments. Studies on using these two techniques for CWA detection have also been conducted. These techniques have very good sensitivity and large linear response dynamic range. They can sense a group of chemicals that can be ionized, and results are the sum of total detectable chemicals in the sample. Because of their nondiscriminative natures, these techniques are nonspecific for CWA and TIC detection. Although these methods can detect some organic CWAs and TICs with sufficient high volatility in the laboratory where the existence of the compounds is known, their nondiscriminative detection characteristics have prevented their usefulness in detecting highly toxic compounds under field conditions. Their main application as a field detector is for identifying volatile organic chemicals (VOCs). A separation method, such as the addition of a GC column ahead of the detection, is necessary to provide these technologies for potential identification of individual substance contained in a sample based on the elution times.

Both techniques measure the current generated by the ionized species sensed by the ion collector. The photoionization technique ionizes molecules using a high-energy photon source, such as ultraviolet (UV) radiation, while flame ionization technique burns organic molecules in a hydrogen flame, similar to the flame photometric detection (FPD) technique described earlier, that ionizes the decomposed fragments. The ions generated are detected similarly for both photoionization detectors (PIDs) and flame ionization detectors (FIDs). By applying an electrical field gradient across the ionization region to drive the ions to the electrodes, the ions release their charges to produce signals that can be processed.

The main differences between these two techniques are their ionization source and mechanism. The photoionization technique is a nondestructive method of detection, while flame ionization method destroys the sampled material.

11.1 PHOTOIONIZATION TECHNIQUE

PID detects volatile organic substances in very low concentrations (parts per billion to parts per million) with a large linear dynamic range of up to six orders of magnitude. Since the first photoionization detector that was manufactured in 1974, the technique has been used in numerous applications to detect and analyze organic airborne substances. This technology uses a high-energy UV radiation source to ionize molecules. The ionized molecules are driven to the ion detector by an electric field gradient. The detector analyzes the electric current generated by ions, which is proportional to compound concentration. Photoionization is an important detection technique used in GC systems for the analysis of organic substances in various samples. Aromatic hydrocarbons or heteroatom-containing compounds (like organosulfur or organophosphorus species) are routinely analyzed with photoionization detectors in the laboratory. PID can also detect certain inorganic compounds that other types of detectors may not. This is because these inorganic species have ionization potentials (IPs) that are within reach of commercially available UV lamps (8.3 to 11.7 eV). PID is nonselective in the sense that any molecules with lower ionization potential can be ionized by an ionization source having higher energy. A UV source can improve selectivity in that the compounds with higher ionization potentials would not be ionized or detected when a lower energy source is used.

11.1.1 Photoionization

As discussed in Chapter 7, when electrons absorb external energy from the hydrogen flame, an atom or a cluster are excited and jump to a higher energy level. The photometric method detects the radiation released by the atom or cluster when it returns to its lower energy level. When the external energy — provided by a UV source in a PID — absorbed by an atom or molecule is sufficient, photons in the radiation can knock off electrons and thus ionize the molecule as illustrated in Figure 11.1. Because ionization is induced by high-energy photons, the process is called photoionization. Photoionization can be expressed as

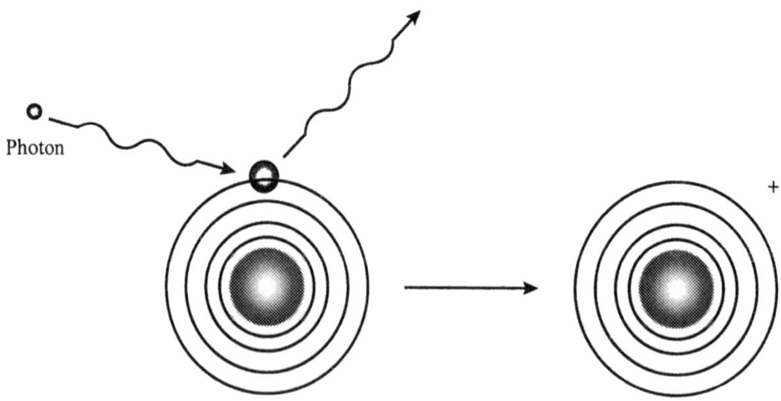

Figure 11.1 Photoionization.

$$R + h\nu \rightarrow R^+ + e^- \tag{11.1}$$

where R represents the species to be ionized, and hν represents the energy of a photon required to ionize species R.

For ionization to occur, the energy (hν) applied must be greater than the energy needed to remove electrons from the species (IP, or ionization energy [IE]). IE is defined as the energy needed to remove an electron from its atomic orbit and place it at an infinite distance. This energy is expressed in units of electron volts (eV). For most compounds, more than one electron can be removed. Removal of second or third electrons, however, requires much higher energy levels than that required for removing the first electron.

11.1.2 Ionization Source

To ionize a molecule by the photoionization process, the energy of a photon must be greater than the ionization potential of that molecule. A higher-energy electromagnetic radiation than the infrared radiation, such as the ultraviolet radiation, can be used as one of the ionization energy sources. Johann Ritter (1776–1810) discovered UV in 1801, which in turn was inspired by the discovery of IR by William Herschel in 1800. Ritter used silver chloride in his experiment and found that the compound turned black much more efficiently when exposed to light just beyond the violet end of the spectrum than when exposed to red light (IR). Ritter called the invisible light he discovered "chemical rays," which is the ultraviolet light or radiation known today. The energy level of UV light ranges from 3 to 10^3 eV. Refer to Table 7.1 for the position of UV radiation relative to other types of radiation in the electromagnetic spectrum.

UV radiation can be generated by different means. For a photoionization detector, a UV lamp is used to provide continuous UV irradiation. Most UV lamps used in photoionization detectors have energy levels between 8.3 eV with wavelengths of 150 nm to 11.7 eV with wavelengths of 106 nm. The energy level increases with shorter wavelength sources. Most photoionization detectors use only one UV lamp at a time. The lamps, with varying eV values, can be exchanged based on application needs.

Air is comprised of N_2 (78%) and O_2 (21%), with relatively small amounts of other substances, including argon, water, and CO_2. IE values for these gases are greater than 12 eV. The IE of the major component of the air, nitrogen, is 15.58 eV. The IE of oxygen is 12.075 eV. Water and CO_2 have the IE values of 12.59 eV and 13.79 eV, respectively. Therefore, UV sources used in common photoionization detectors cannot ionize them. This eliminates them as potential interfering compounds for photoionization detection and analysis. However, water vapor in the atmosphere has been found to cause significant effects on PID detection instruments. This effect is, perhaps, the result of factors other than the ionization effect because the ionization potential of water is higher than for the UV lamp.

11.1.3 Ion Detection

Sample molecules are ionized by UV radiation in the sample chamber. A pair of collection electrodes is used to apply electric potential to force the formed ions to move toward the electrodes. When the ions contact the electrodes, they become neutralized by releasing their charge and generate an electric current that can be detected by an electronic device. The electric current generated by the ions is proportional to the amount of ionized substances that enter the detector. Therefore, the concentration of molecules in the sample could be determined by measuring the current.

11.1.4 Instrumentation

Figure 11.2 is a diagram of a typical photoionization detector. In principle, a PID includes a sample inlet, ionization source, ion collector, and sample outlet. Vapor samples are introduced through the sample inlet system, which includes a GC column for general GC applications. In the field, the sample vapor is generally drawn into the detector without gas chromatographic capability. The use of a GC column increases detector selectivity, but also slows down response time. The analysis is performed continually while the sample is drawn through the detector. When the sample enters the ionization area where the electric field is applied, molecules in the sample are ionized if their IP values are lower than the UV-source energy level. The ions generated are driven to the electrodes where the ions release their charges to produce the electrical signal read by the detector. After the ions release their charges, the particles revert to their original state and exit the detector. Therefore, the photoionization method is nondestructive; the exhausted sample can be further analyzed using other methods as necessary.

11.1.5 Technique Specification

Theoretically, any substances that can be ionized by the selected UV source can be detected. In the past, this detection technology was used to detect various organic substances., but it is also effective in detecting inorganic substances (e.g., H_2S, NH_3, etc.). PID is quite sensitive, as its limit of detection reaches low parts-per-million levels and into parts-per-billion levels. Response dynamic range can be up to six orders of magnitude.

A disadvantage of the PID technique is that when a separation technology is not used (e.g., GC), the signal is generated by all ionizable chemicals in a sample rather than by a single targeted compound. All chemical components in the vapor that can be ionized contribute to the signal. Thus, the response may not represent the targeted compound unless it is the only compound that could be ionized. For example, when an 11.7-eV UV lamp is used, all chemicals with an ionization potential lower than 11.7 eV will contribute to the signal, and the result is the sum of all ionized components. Similarly, all chemicals with an IE lower than 9.5 eV will contribute to the signal when a 9.5-eV lamp is used (Figure 11.3). In brief, photoionization detectors are not selective.

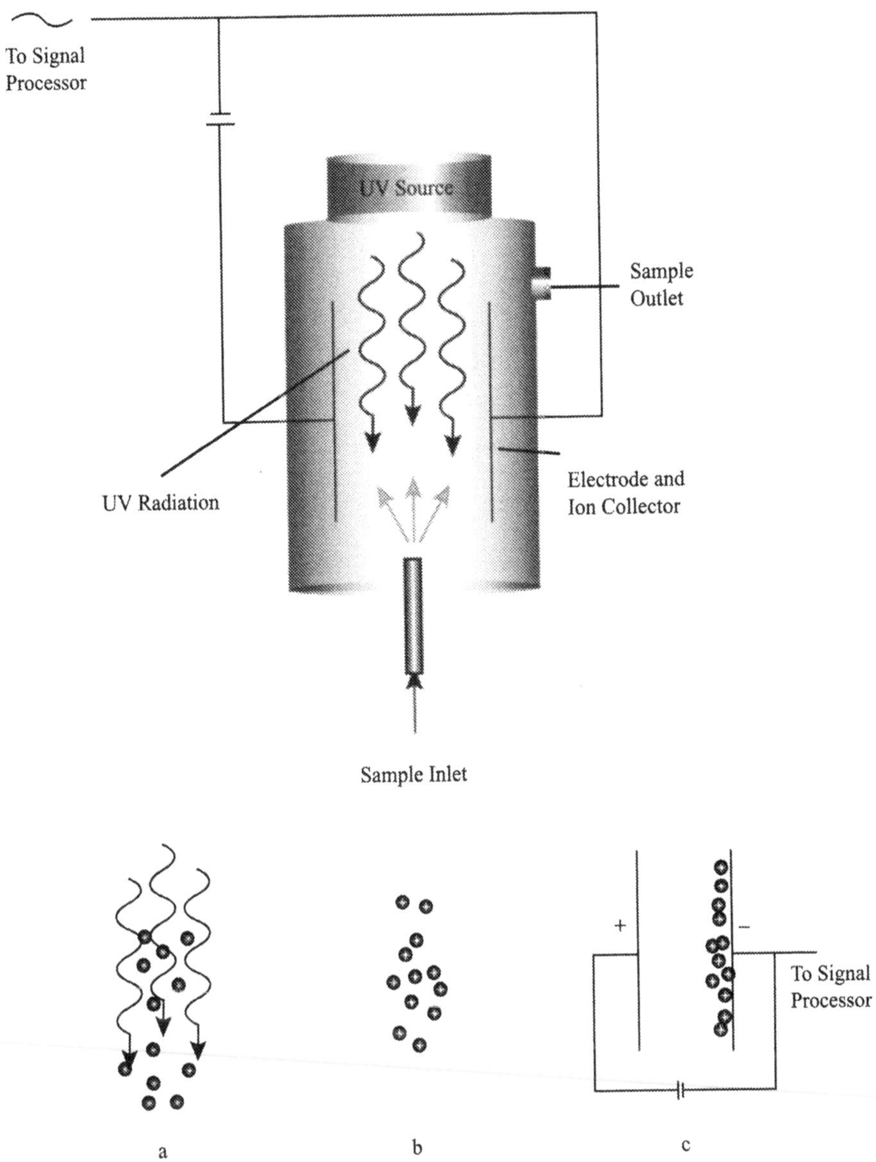

Figure 11.2 Schematic diagram of PID.

Another characteristic feature of PID is that for organic matter that can be ionized, its response is roughly proportional to the number of carbon atoms in the molecules. At the same mole concentration, the signal generated by molecules containing six carbon atoms will be roughly double that generated by molecules containing three carbon atoms. For this reason, PID is also called the "carbon counter" method.

Although photoionization is not selective, by using a lower-energy UV source, fewer chemicals will be detected since those chemicals with a higher IP will not be

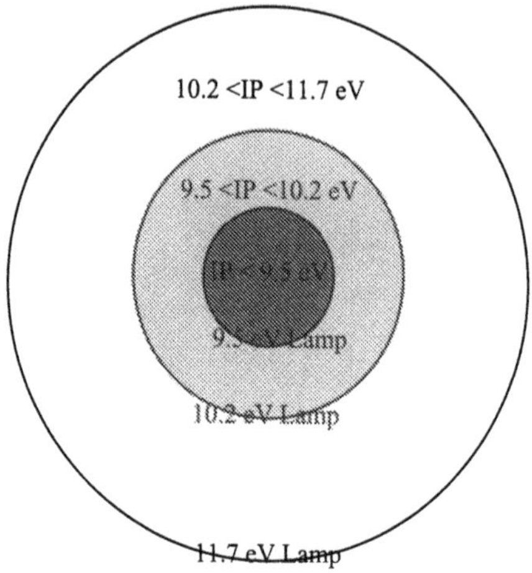

Figure 11.3 Ionization energy and energy UV lamp.

ionized. Interference is therefore reduced, but not eliminated. To positively identify a chemical using the photoionization detector, a chemical separation method such as a GC column must be used before the sample enters the detector.

11.2 FLAME IONIZATION TECHNOLOGY

Flame ionization, a common detection and analysis technology for volatile organic chemicals, is similar to photoionization. FID uses a hydrogen flame rather than UV radiation as the means to produce ions from decomposed organic fragments in a vapor sample. This technique also has a very good limit of detection (parts-per-billion to low parts-per-million level) for VOCs with a large linear dynamic response range up to seven orders of magnitude. It is a nonselective method that responds to any molecule containing carbon–hydrogen bonds without the ability to identify the various compounds. Consequently, FID cannot effectively detect inorganic compounds such as H_2S and NH_3. Because of its nonselective nature, it is more useful as a type of GC detector than as a field detection device for toxic compounds. FID has been incorporated in handheld field instruments for VOC detection when precise identification is not a requirement.

FID is similar to the flame photometric technique except in the ways that signals are harvested. For the flame photometric technique, the hydrogen flame temperature is between 2000°C and 2500°C. At this temperature, all organic substances decompose and form ions containing carbon-hydrogen. The actual formation of ions in the hydrogen–air flame is a rather complex process. Basically, the organic molecules undergo a series of reactions including thermal fragmentation, chemi-ionization, and ionic molecule and free radical reactions to produce the charged species. The quantity of ions produced is proportional to the number of reduced carbon atoms present in the flame and hence the number of molecules. The FID responds to the number of carbon atoms entering the detector per unit of time, which is similar to photoionization detectors. For this reason, FID is also referred to as a "carbon counter" detector.

All organic matter decomposes when burned in a hydrogen flame, and ions are generated for many organic molecules. The FID sums all chemicals contained in the sample that generated the ions in the hydrogen flame. Flame ionization detectors are insensitive toward noncombustible gases such as H_2O, CO_2, SO_2, and NO_x. Therefore, these substances do not interfere with detection. Consequently, FID is quite often is coupled with a photoionization device to identify VOCs.

The diagram in Figure 11.4 shows the operational principle of FIDs. FIDs, constructed similarly to FPDs, are comprised of a sample inlet, hydrogen fuel inlet, combustion chamber, and electrodes that provide the electric field and act as an ion collector. The decomposed sample is vented through a chimney exhaust. The major difference between an FID and a FPD is the method used to process the detection signal. Instead of measuring light emission from excited ions, the FID measures electric current generated by the ions.

Samples can be introduced into an FID either directly or through a GC column. Hydrogen, air, and sample vapor are mixed in the combustion chamber and burned. Because of the high temperature of the hydrogen-rich flame, organic substances in the vapor are decomposed to hydrocarbon fragments and ionized. The ions move along the electrical field toward the electrodes where current is produced when the ions hit them. This signal is then sent to the electronic signal processor.

There are a few disadvantages in using photoionization and/or flame ionization instruments for field CWA and TIC detection. Both PIDs and FIDs are not "selective": they respond to all chemicals that can generate ions via the UV source (in PID) or hydrogen–air flame (FID). Therefore, no specific identification and corresponding concentration can be determined based solely on PID or FID detection responses. Separation of sample compounds is necessary for both technologies to identify target chemicals or determine respective concentrations. Another concern about the FID is the use of hydrogen gas as fuel. As discussed in Chapter 7, a hydrogen cartridge is an alternative to a compressed gas cylinder. Nevertheless, the use of hydrogen still creates logistic issues.

11.3 COMPARISON OF PHOTOIONIZATION, FLAME IONIZATION, AND FLAME PHOTOMETRIC TECHNIQUES

PIDs and FIDs have similar detection capability for VOCs in that the ionization process and signal collection of both are nonselective. Although very useful as VOC detectors, they are quite limited when used as field detectors for CWAs. Detection of CWAs can occur only through the use of response factors equivalent to respective calibration gas. Methane gas and isobutylene are used by FID and PID, respectively, as the reference calibration gas. Response factors correlate detector responses calibrated against the reference gases to a known concentration of a given compound. The usefulness of response factors, unfortunately, is valid only if the sample contained the targeted chemical without any other influences (i.e. in a known situation). They are not viable CWA or TIC detectors because they do not provide specific

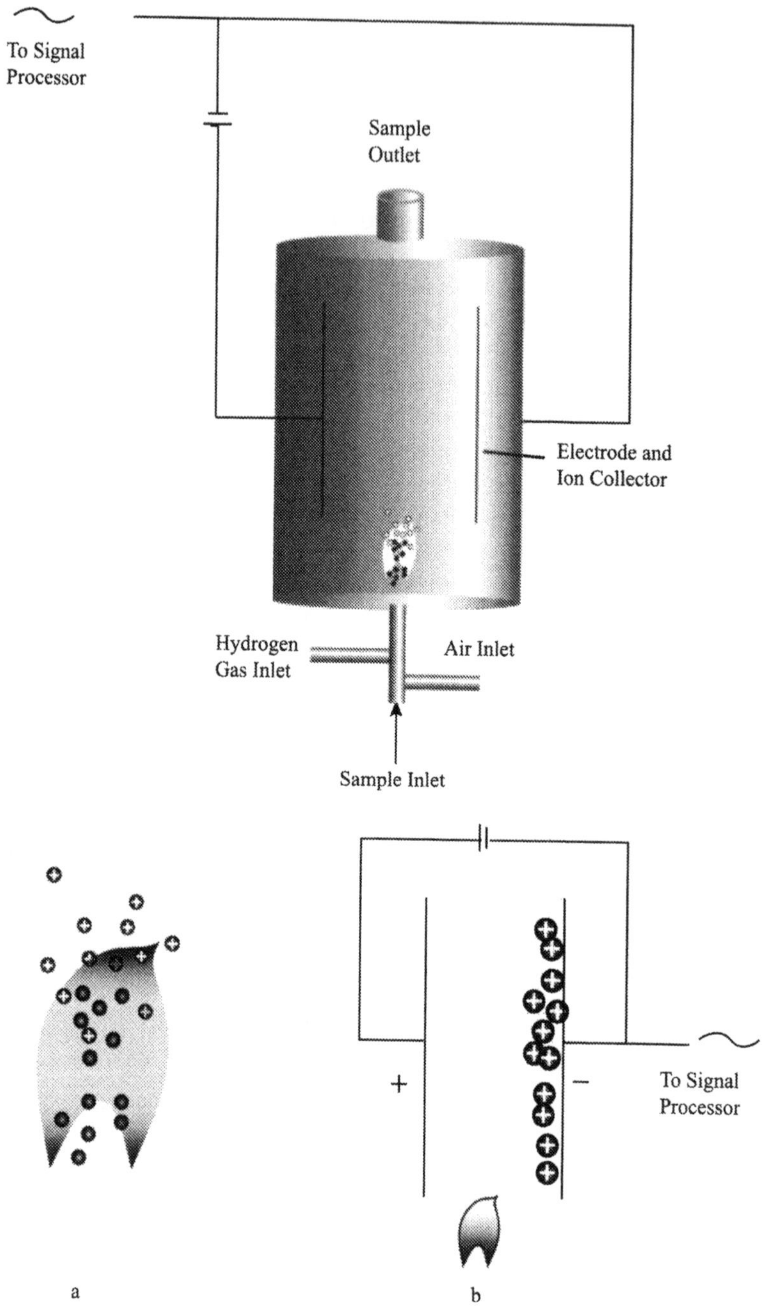

Figure 11.4 Schematic diagram of FID.

chemical concentration data. Given that the only difference between FID and FPD is the way that signals are processed, FPD has the advantage of greater selectivity. Besides the detection of phosphorous and sulfur components as used in general CWAs, FPDs can similarly detect other element emissions through the use of different light filters. The ability to selectively detect target chemical light emission makes the FPD more selective. It detects those chemicals containing elements that emit light upon burning in the hydrogen–air flame. Theoretically, other chemicals lacking these elements will not interfere with the detection. Thus, FID and PID are excellent VOC detectors, and multiple compounds can be simultaneously and quickly detected. Their use as CWA and TIC detectors requires component separation of the sample before detection. For this reason, the PID and FID are not being aggressively pursued as CWA detection devices, although many FID and PID manufacturers would like them to be. FPD offers much more specificity in CWA and TIC detection.

11.4 APPLICATIONS

The Foxboro TVA-1000B Toxic Vapor Analyzer® is a dual PID and FID system. The two detectors can be used individually or in combination. Results are displayed in parts per million, parts per billion, or as percent concentration. The PID has a dynamic range of 0.5 to 2000 ppm of isobutylene using a 10.6-eV lamp, while the FID has a dynamic range of 1 to 10,000 ppm of methane. A pressurized hydrogen cylinder is used to provide the hydrogen for FID operation. The 85-cc cylinder of compressed hydrogen at 2200 psi lasts for approximately 8 hr of continuous operation.

This detector was evaluated under the Domestic Preparedness Program (DPP) (http://hld.sbccom.army.mil/ip/reports.htm#detectors) as a potential viable CWA detector. Response factors for diverse CWAs were created against known concentrations, under known environmental conditions. Evaluators concluded that this detection mechanism would be susceptible to interference. Its nondiscriminatory detection results cannot assure users whether they are caused by a CWA or other contaminants that may coexist in the air.

The DPP has also evaluated PID detectors made by RAE Corp. and Mine Safety's Passport II®, and the FID manufactured by PerkinElmer Corp. called MicroFID®. These detectors were considered too nonspecific to be considered as viable detection devices to provide adequate warning and information about the existence of CWAs or TICs. Identification of toxic substances is essential if these technologies are to be adopted. Short descriptions of available FID and PID instruments follow.

The Passport PID II Organic Vapor Monitor (Mine Safety), ppbRae® (Rae Systems), and HNμ Model PI-101 Trace Gas Analyzer® (HNμ Systems, Inc.) (Figure 11.5) are all PID detectors. They use UV lamps at 10.6 or 11.7 eV to ionize molecules and detect the presence of the photoionized species in the sample stream. They do not distinguish among individual chemicals. The concentration displayed represents the concentration of all ionized species. Moisture, oxygen, methane, and other compounds will quench the PID signal to cause underreporting of concentration readings. A daily calibration check is part of the setup routine for operation.

The Passport PID II Organic Vapor Monitor weighs 2.5 pounds with the battery attached, with dimensions of 8.125 in. × 2.875 in. × 3.75 in. It draws in 350 cc/min, and is said to be operable between 0°C to 40°C with a noncondensing of 1% to 95% relative humidity (RH).

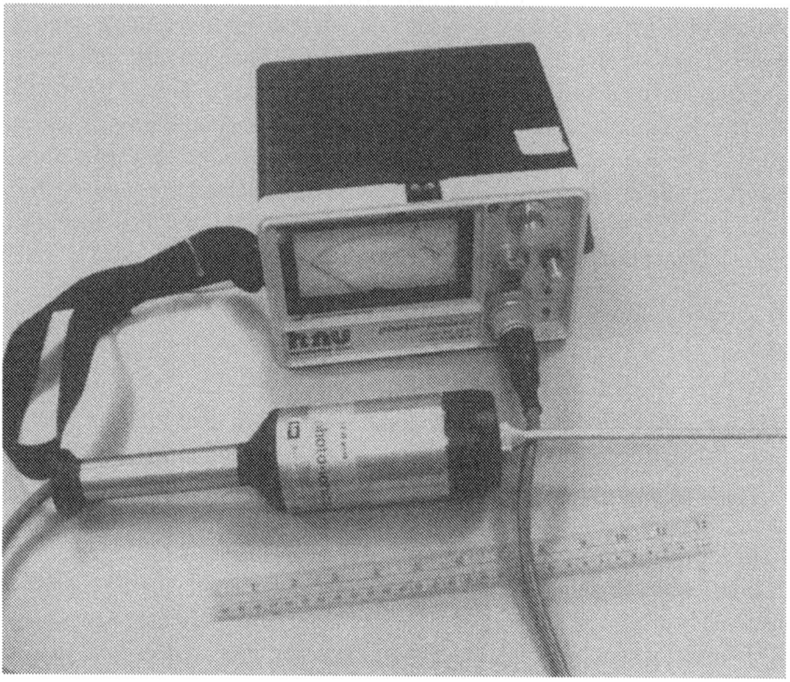

Figure 11.5 Photoionizer detector. Photo courtesy of K. Y. Ong.

HNu Model PI-101 Trace Gas Analyzer is a portable instrument that can detect, measure, and provide a direct reading of the concentration. The analyzer consists of a sample probe, readout assembly, and battery charger. The sample probe contains sensing and amplifying circuitry, and the readout box contains meters, controls, and power supply. The standard probe uses a 10.2-eV lamp and has a response time of less than 5 sec to reach 90% of full detection capacity. It has several sensitivity scales and a detection range from 0.1 to 2000 ppm. It is operable up to 90% RH at temperatures ranging from −10 to 40°C. The detector weighs approximately 9 pounds with the probe attached, and measures 8.25 in. × 5.25 in. × 9.5 in. The test conducted at the U.S. Army Soldier Biological and Chemical Command (SBCCOM) used the 11.7-eV probe to provide maximum ionization potential for assessment of its capability to detect CWAs.

RAE Systems Inc. of Sunnyvale, California is the manufacturer of the parts-per-billion Volatile Organic Compound (VOC) Monitor (ppbRAE Model PGM-7240) (Figure 11.6). According to the manufacturer, the ppbRAE is a PID with a standard 10.6-eV or optional 9.8- or 11.7-eV UV lamp. The ppbRAE is a durable, lightweight (19.5 oz, including battery pack), handheld detector designed for continuous monitoring of dangerous environments for VOCs. Alarm levels can be preset for low, high, short-term exposure limit (STEL) and time weighted average (TWA) levels. There is an audible alarm as well as a red flashing LED, along with the direct parts-per-billion display readout. The detector also has point data logging capability for postevent downloading to a computer.

Figure 11.6 ppbRAE detector. Photo courtesy of K. Y. Ong.

The ppbRAE uses a dual-channel PID and an electrode-less discharge UV lamp as the high-energy photon source. The built-in sample pump draws in the vapor sample at a nominal flow rate of 400 cc/min. The sample passes by the UV lamp where it is photoionized, enabling the ions to be detected as current is generated. The instrument uses the sensor readings to calculate the gas concentrations based on a known response factor derived from a reference calibration gas. The ppbRAE was evaluated in its "Hygiene" mode, where the monitor runs continuously and the liquid crystal display displays the instantaneous readings. Operational temperature range is −10°C to 40°C and RH range is 0% to 95%.

These PID detectors exhibited wide ranges of response factor (RF) values between units as well as within the same unit. RFs represent the ratio between the signals obtained from exposures to the targeted substances and the calibration gases used to span the detection range with known concentration. They displayed varied and inconsistent parts-per-billion values when exposed to similar concentrations of CWA vapors. Consequently, no meaningful RF versus agent concentration relationship could be determined. Residual effects from CWA challenges and/or calibration drift were observed. The UV lamps, used to ionize the vapor samples for detection, are easily contaminated by exposure to dust, dirt, moisture, and other compounds. Frequent and thorough cleanings are required to maintain detector performance.

PerkinElmer Corp. produces the Photovac MicroFID® Handheld Flame Ionization Detector (Figure 11.7). This detector was selected for testing based on a survey of existing detection devices by Battelle Memorial Institute. The survey identified the detectors most likely to be used by local responders in the event of a terrorist incident involving CWA(s). No attempt was made to optimize chemical agent detection capability. No pretest theoretical assessment was made on the detectors except to learn operating procedures from the manufacturer's user manual.

Figure 11.7 MicroFID detector. Photo courtesy of K. Y. Ong.

Unpredictable CWA detection performance prevented the establishment of a reliable response curve. Test results suggested that the MicroFID in its current configuration could not be used effectively for CWA detection. Recalibration checks clearly suggested that detector sensitivity was not degraded by exposure to CWAs, as methane detection capability was nearly constant. Methane detection responses also did not appear to be affected by RH changes. However, CWA detection at varied humidity conditions showed that humidity grossly affected instrument sensitivity. Testing was discontinued after ambient-temperature agent sensitivity tests because of poor performance in CWA detection.

Based on observations from testing of various devices from different manufacturers, both PID and FID instruments would not be the detectors of choice for CWA or TIC detection without the addition of a GC type of compound separator to increase selectivity. In brief, other detection technologies are preferable. (The related DPP test report is posted at http:hld.sbccom.army.mil/ip/reports.htm#detectors.)

11.5 FACT SHEETS ON SELECTED PHOTOIONIZATION AND FLAME IONIZATION DECTECTORS

Detector Name

Foxboro TVA 1000B PID Detector

Manufacturer/Distributor and Contact Information

The Foxboro Company
33 Commercial Street
Foxboro, MA 02035
Tel: 508-543-8750

Technique Description

The TVA 100B PID Detector is man portable detector utilizes 10.6 eV UV source for photoionization. It co-existed with a companion FID for broader detection capability. The two complement each other.

Chemical Detection Capability and Performance

It can detect nerve, blister, blood, and choking agents as well as many TICs that could be ionized with near real time detection. Unfortunately, the detections are non-specific. Responses cannot be attribute to a certain compound in the unknown situations, therefore, cannot be readily used for toxic environments. It can only serve as an indication that something (VOC) is in the air.

Other Feature

Audio and visual alarm indications. Response factor can be determined for laboratory uses. Response factor varies with moisture contents in the sample, in general, due to quenching effect.

Domestic Preparedness Test Report

Testing of Commercially Available Detectors Against Chemical Warfare Agents: Summary Report, February 1999. (http://hld.sbccom.army.mil/downloads/reports/detectors_summary.pdf)

Detector Name

MiniRae Classic, ppbRae, MiniRAE 2000

Manufacturer/Distributor and Contact Information

Rae Systems, Inc
1339 Moffett Park Dr.
Sunnyvale, CA 94089
Telephone: 877-723-2878

Technique Description

The MiniRae Classic, ppbRae, MiniRAE 2000 detector are handheld portable detector utilizes 10.6ev UV lamp for photo-ionization ionization.

Chemical Detection Capability and Performance

It can detect many TICs and possible some CWAs that could be ionized with near real time detection. Unfortunately, the detections are non-specific. Responses cannot be attribute to a certain compound in the unknown situations, therefore, cannot be readily used for toxic environments. It can only serve as an indication that something (VOC) is in the air.

Other Feature

It has audio and visual alarm indications and data logging features. It can be networked. Response factor can be determined for laboratory uses. Response factor varies with moisture contents in the sample, in general, due to quenching effect

Domestic Preparedness Test Report

Testing of Commercially Available Detectors Against Chemical Warfare Agents: Summary Report, February 1999.
(http://hld.sbccom.army.mil/downloads/reports/detectors_summary.pdf)

Detector Name

MSA Passport II Organic Vapor Monitor

Manufacturer/Distributor and Contact Information

Mine Safety Appliances
P. O. Box 426
Pittsburg, PA 15230
Telephone: 800-MSA-2222

Technique Description

The MSA Passport II Organic Vapor Monitor is a handheld portable detector utilizes 10.6 eV UV lamp for photo-ionization ionization.

Chemical Detection Capability and Performance

It can detect many TICs and possible some CWAs that could be ionized with near real time detection. Unfortunately, the detections are non-specific. Responses cannot be attribute to a certain compound in the unknown situations, therefore, cannot be readily used for toxic environments. It can only serve as an indication that something (VOC) is in the air.

Other Feature

It has audio and visual alarm indications and can be upgraded to detect new compounds through changeable data libraries. No expendables except for occasional filter changes. It meets MIL standards and currently fielded by many nations. It can be networked.

Domestic Preparedness Test Report

Testing of Commercially Available Detectors Against Chemical Warfare Agents: Summary Report, February 1999.
(http://hld.sbccom.army.mil/downloads/reports/detectors_summary.pdf)

Detector Name

HNu PI 101 and DL102 Organic Vapor Monitor

Manufacturer/Distributor and Contact Information

PID Analyzers, LLC
25 Walpole Park South Drive
Walpole, MA 02081
Telephone: 800-724-5600

Technique Description

Both detectors are photo-ionization based using several different UV lamps of 9.5, 10.2, 10.6, and 11.7eV as interchangeable modules for different applications

Chemical Detection Capability and Performance

These detectors are used primarily for VOC detection and can detect many TICs and possible some CWAs that could be ionized with near real time detection. Unfortunately, the detections are non-specific. The response is fast but cannot be attribute to a certain compound in the unknown situations, therefore, cannot be readily used for toxic environments. It can only serve as an indication that something (VOC) is in the air.

Other Feature

The DL 102 is a more advanced model from model 101. They has builtt-in library of sensitivitiesa dn 7000 points data logging capability.

Domestic Preparedness Test Report

Testing of Commercially Available Detectors Against Chemical Warfare Agents: Summary Report, February 1999. (http://hld.sbccom.army.mil/downloads/reports/detectors_summary.pdf)

Detector Name

Perkin Elmer MicroFID Flame Ionization Detector

Manufacturer/Distributor and Contact Information

The Perkin Elmer Corporation
761 Main Avenue
Norwalk, Connecticut 06859
Telephone: 203-761-5330

Technique Description

This is a handheld portable flame ionization based detector using a small compressed cylinder of hydrogen. It is aimed for detection of VOC that will burn. It does not distinguish between individual pollutants.

Chemical Detection Capability and Performance

The detector is used primarily for VOC detection and can detect many TICs and possible some CWAs that could be ionized with near real time detection. Unfortunately, the detections are non-specific. Responses is fast but cannot be attribute to a certain compound in the unknown situations, therefore, cannot be readily used for toxic environments. It can only serve as an indication that something (VOC) is in the air.

Other Feature

It has data logging capability. One fully charged hydrogen cylinder lasts approximately 11 hours of operation.

Domestic Preparedness Test Report

Domestic Preparedness Program: Testing Of Photovac Microfid Handheld Flame Ionization Detectors Against Chemical Warfare Agents - Summary Report, October 1999 (http://hld.sbccom.army.mil/downloads/reports/photovac_microfid_summary_report.pdf)

CHAPTER 12

Future Trends in CWA and TIC Detection

We have discussed various technologies employed for detection of CWA and TIC vapors. Each of those technologies offers unique desirable as well as undesirable features. There is considerable room for improvement. Increasing detection reliability and reducing the frequency of false responses continue to be major challenges. The recent emphasis on TIC detection capability in addition to that of CWAs has also created substantial challenges for researchers and developers. Such requirements limit the number of technologies that can be effectively used.

For example, current IMS detectors, in theory, could be made to detect diverse TIC compounds in addition to their CWA detection capability. IMS detectors attempt to identify compounds based on compound characteristics to form peaks in the mobility spectrum. Each of the targeted compounds can then be assigned a "window" or "electronic gate" to identify the substance when the instrument detects a peak in that region. Drift time varies with respect to temperature or pressure variations. Therefore, an excessively narrow drift-time "window" for a given compound would prevent the instrument from detecting it at temperature extremes.

Length of the drift region is also a limiting factor. The number of segments available for window assignments determines the number of species that can be identified. As discussed previously, many substances have a similar drift time. If these fall into respective gates in sufficient quantity, the alarm is triggered. Therefore, the larger the number of species to be detected, the higher the frequency of false alarms for many substances. IMS detectors would need to contain libraries geared to detecting more specific groups of compounds in diverse operational scenarios. Perhaps with the current advances in microelectronics and software engineering, a single IMS detector could contain several resident detection libraries, and each of them could be searched to identify and list suspected compound(s) correlated to the detected peak(s) prior to sounding the alarm.

Besides the use of IMS technology, researchers are concentrating on the use of polymer coatings on microscaled sensors so that many of these sensors could be arranged in an array to detect diverse compounds. Other techniques, such as chemical resistance, surface acoustic waves, light emission, semiconductor cells, and fiber optics, are being used to harvest responses from microsensor arrays. Polymer coatings can be made quite specific. Each coating could be made to carry out a reversible chemical reaction for a specific family of compounds. Detection results from combining inputs from sensor arrays and developing pattern recognition algorithms. These approaches are still in their infancy, however. Much improvement is required to harden polymers to withstand temperature extremes and vibration stress, increase longevity, and make possible fast reversibility without residual effects.

Many researchers are attempting to integrate currently available detection technologies. It was hoped that the CWA or TIC detection capabilities of multiple sensors could be integrated, data analyzed through computer programming, before deciding whether a detected substance is indeed the targeted CWA or TIC, and not interference from nontoxic components of the sample. These approaches are aimed at minimizing the potential for false alarms and producing more reliable detection when the sample indeed contains toxic compounds.

The Challenges for the Chemical Sciences in the 21st Century Workshop on National Security and Homeland Defense brought together chemical scientists and engineers from academia, government, national laboratories, and industrial research laboratories to provide a broad range of experiences and perspectives needed to fight the war against terrorism. Their presentations and discussions showed that chemists and chemical engineers have the skills to develop solutions for national security and homeland defense problems. The conference showed that advances in science and technology would be much more promising when chemists and engineers and related disciplines collaborate. The organizing committee derived two multifaceted findings from workshop discussions.

1. Extensive research opportunities exist throughout the many facets of chemistry and chemical engineering that can assist national security and homeland defense efforts.
2. Cooperative research in systems analysis is required to apply the new knowledge to developing tools and products required for national security and homeland defense.

Three major needs identified at the workshop follow:

1. Detectors with the broadest possible range for chemical and biological agents. New sensing and sampling technologies are needed to yield improved threat detection and identification.
2. More specific detectors that can provide speed, high sensitivity, and reliability.
3. Detection and identification equipment miniaturization for field use. This is the most urgent feature required by soldiers, first responders, and other users. Miniaturized devices will lighten the loads that these users must carry during an incident.

Proper post-incident management is most critical to minimize casualty. Sensors with significantly improved specificity and sensitivity beyond currently available devices can assist authorities in assessing the extent of contamination for effective evacuation and determining when a site is safe for return to normal functions.

Consequently, various subsystems will be needed to support the development of these more sensitive instruments. Systems for reliable sample collection, sample processing, and presentation of chemicals to sensors are essential. Standardized methodologies are needed. Systematic quality assurance on sensor evaluation can only be achieved through standardized methodologies that have proven successful. Toward this end, the National Institute of Standard Technologies is in the process of generating standard testing protocols under which all future detection devices shall be tested for certification.

The U.S. government has drastically increased financial support of research of these sensors, mainly through the Homeland Defense program, DoD, National Science Foundation, and Department of Energy. Private industry is also heavily involved in developing sensors. New sensors or technology improvements are surfacing in the market fairly rapidly. However, as of this writing, none of the current technologies has had any significant impact on emergency preparedness. For sensors to be effectively implemented, they must be highly reliable, fairly inexpensive, widely deployable, and have networking capability.

Development of sensors that can detect and identify the release of toxic materials under various environmental conditions must continue. Effective response to a given agent involved in a chemical attack can only be achieved through the right choice of sensors for the occasion. Effective postrelease management is defined in terms of orderly evacuation to minimize casualties.

As mentioned in Chapter 7, the handheld Fast GC-PFPD may prove to be one of the more reliable detection devices for identifying CWAs and the most toxic TICs. The instrument combines features of concentrating the sample, thermal desorption, gas chromatographic separation, and subsequent pulsed flame photometric detection. The system has been tested, demonstrated, and documented in reports by the inventor using CWA simulants, but has not been tested with actual CWAs because, unfortunately, it has not been commercially available. According to the information obtained from the inventor's web page (www.tau.ac.il/chemistry/amirav), the system is ready for laboratory demonstrations with some external electronics because the prototype unit has not been made and the system has not been commercialized. No CWA test results are available until funding support is available. Therefore, a program with sustained funding to focus and coordinate research and development on promising sensors and sensor networks with an emphasis on fielding the system is needed. Emphasis on finding commercial off-the-shelf (COTS) should be expanded to prototyping highly promising devices as well.

Sensors and detectors require different technologies for different applications. To develop robust sensors, a multidisciplinary systems approach must be taken. Experimentalists, statisticians, engineers, and data analysts must become involved from concept development to fielding of the sensor. Well-designed experiments incorporating appropriate statistics must be used in developing field-worthy sensors to minimize testing costs. Actual and potential interference must be identified and dealt with either through hardware design, use of multiple sensor types, application of multivariate techniques, or better software development. Sensor calibration and drift compensation need to be verified through reliable testing methodologies that can provide reproducible results so that compensation can be correlated and corrected, either mathematically or through hardware improvements.

Once developed, detectors and sensors need to be tested according to strict criteria. Test standards, including LODs and identification of potential problem areas, must be revealed. The U.S. government periodically offers the opportunity for realistic field testing. Wind tunnel testing is available at several sites. Developers should take advantage of these resources for technology validation. It would be useful to develop new, integrated multisource databases to create chemical and biological agent reference libraries. There is a need for consolidation, as various libraries exist among many individual agencies.

It is extremely important to note that false negatives, for whatever reason, from a detector could prove to be fatal if the device cannot detect targeted toxic substances. Therefore, research emphasis should not only focus on reducing the false positive alarm rate. The potential for false negatives is more important and must be addressed. While false positives are nuisance factors, false negatives could lead to fatalities.

In summary, the world has changed considerably since the September 11, 2001, incidents. The term "weapons of mass destruction" or WMD has become a household word. Among WMDs, nuclear, biological, and/or chemical weapons are among the highest-ranked possibilities for terrorist groups to deploy besides those used in the 2001 incidents. We have limited this book to the discussion of topics related to chemical weapons.

This book was not intended to cover everything about chemical weapons. There are many related articles, books, publications, and government reports that discuss and summarize various detection technologies, as well as describing CWAs and TICs. The book is a condensation of years of hands-on experience with CWA substances and various detection technologies. Our goal has been to present the subject matter in as simple a form as possible so that the information can reach a large spectrum of readers. We believe that should an incident occur, if people have general knowledge about CWAs and TICs, the panicky response — which typically leads to a much larger number of casualties than would occur without panic — will be minimized. We also hope that this book will provide decision makers with information on the various detection technologies and devices that will help them make decisions on selecting, evaluating, and improving devices for their specific needs.

TICs and CWAs remain the preferred WMD for terrorist attacks. Although these chemicals do not cause widespread casualties and destruction on the level of biological agents or nuclear weapons, they are relatively easy to obtain, conceal, and disseminate, and thus constitute a serious threat. This is especially true for urban areas where there are large aggregations of people in big buildings, subway systems, airports, and sport events. Release of toxic chemicals in a public place could easily occur by an individual or small group of terrorists who are willing to sacrifice their lives. The deployment of toxic chemicals may even permit the disseminator(s) the opportunity to escape, when compared to the use of explosives in suicide bomber incidents. Once the chemical is set for release, the terrorist has a better chance of exiting the area before the innocent victims as the concentration of toxic chemical vapor silently increases. A few hundred grams of nerve agent is sufficient to cause massive casualties and disruption. Nerve agents can easily be concealed, such as in a soda or water bottle.

Toxic chemicals, flammable materials, and explosive materials are readily available for terrorist groups to cause massive casualties with high visibility to create public panic. The 1995 release of sarin (GB) gas in the Tokyo subway system is a prime example of what, when, where, and how toxic chemicals could be deployed among civilians to gain worldwide attention.

Although CWAs have existed for many years, and instruments for detecting the more lethal agents have been developed over the years, the perfect detection device is still unavailable. To expect a single device to provide adequate detection of all lethal agents as well as TIC compounds would be naive. Each of the technologies has pros and cons. The Domestic Preparedness Detector Testing Program that tested many commercially available, off-the-shelf (COTS) devices has revealed many deficiencies of existing detection devices. We were fortunate to have been actively involved in many of these tests to gain firsthand knowledge of device characteristics.

The Domestic Preparedness Program has evaluated most of the modern COTS detectors that may have the capability to detect CWAs with reliability and sensitivity. The program solicited detector candidates from private sector manufacturers and did not receive many responses, which indicates the current limited potential of finding additional suitable CWA detectors in COTS scenarios at this time.

Since the September 11, 2001, incidents, the U.S. has formed the Homeland Defense Department to combat terrorism. Substantial funding is committed to various disciplines to research and develop better detection devices. Perhaps more reliable devices may be developed soon. The emphasis has been on developing inexpensive devices so that they could be widely deployed. Unfortunately, this emphasis has prevented or hindered further enhancement of highly promising technologies that are considered "too expensive" to produce and are thus not funded. The fragmented approach of seeking an ideal detector through the COTS route may not be fruitful. Individual researchers or manufacturers do not have the resources to develop complex systems that could detect various toxic materials and still be "inexpensive." It would be ideal, of course, if reliable detectors could be had cheaply. But, we must ask ourselves, is it possible? Until we reach that stage, it is hoped that decision makers in various government programs will expand their horizons to first fully develop promising devices that have the potential to become reliable detectors, in spite of higher price tags, while also continuing to seek better alternatives.

The recent trend for data fusion of more than one technology in a package is an excellent start. Data fusion of several technologies will, undoubtedly, result in an expensive system. For that to be successful, however, experts in each of the technology disciplines must come together with open minds to work cooperatively. Manufacturers must be willing and open-minded enough to share their proprietary knowledge and work toward the common goal. Reluctant to accept or incorporate each other's merits is a roadblock to success. It is our hope that the respective merits of various devices can be combined to achieve the ideal.

No matter how sensitive and reliable a detection system can be, it will detect nothing (and warn no one) until a target toxic chemical has been released. The best that a detection system can do is to provide rapid and advance warning at a distance from the release before the concentration reaches harmful levels so that people can escape the area before excessive exposure causes catastrophic consequences. Reliable detection devices are vital for the postincident management.

We hope this book will serve as a useful reference to those in the field of developing chemical detectors. Device testing is a vital process toward successful development. It is also hoped that this book will provide the general public with an overview of chemical warfare agents.

APPENDIX A

Material Safety Data Sheet

The Material Safety Data Sheet (MSDS) of a chemical provides the necessary and pertinent information for its safe handling. Federal regulations mandate that an MSDS for each chemical used in a workplace be available and accessible to workers. An MSDS produced by the manufacturer or the vendor is usually shipped with a chemical. In general, MSDSs can be found at manufacturers' websites. If an MSDS for a particular chemical is not already available, individuals or companies using the chemical should contact the supplier.

The MSDS provides information on physical and chemical characteristics; potential health symptoms and effects in case of exposure; how the chemical should be stored; safety precautions on handling and working with the chemical; and steps to follow in case of an accident. It describes the possible physical, health and environmental hazards involved when working with the chemical. An MSDS is not only useful for personnel who work with the chemical, but also for first responders in an incident involving the chemical. The data sheet contains instructions on what to do when exposure occurs, and how to recognize symptoms of overexposure that help first responders to handle the incident properly. By following instructions in an MSDS, the potential for an accident to occur and the hazards it may cause can be minimized.

The MSDS for a given chemical or compound contains many sections, and formats differ by manufacturer. However, all MSDSs for a particular chemical will include similar information. A typical MSDS includes the following:

- Product and company identification
- Composition/information on ingredients
- Hazards identification
- First-aid measures (or health hazards)
- Fire-fighting measures (or physical hazards)
- Accidental release measures
- Handling and storage

- Exposure controls/personal protection
- Physical and chemical properties
- Stability and reactivity
- Toxicological information
- Ecological information
- Disposal considerations
- Transport information
- Regulatory information
- Other information

In a typical MSDS, all of the above would appear as section headings. Appearing below is a hypothetical example of an MSDS: the right column presents example data, and the left column explains what should and could be found in each section, as well as term definitions. Note: The following is *not* a real-world MSDS for the example compound.

Material Safety Data Sheet

Content	Example
Section 1: Chemical Product and Company	
This section identifies the chemical by name, vendor's catalog number, synonyms, formula, and molecular weight, and so on. It may also contain descriptive terms to further help identify the material, such as its chemical family. The manufacturer's name, address, telephone number, and emergency telephone number are found in this section.	MSDS Name: Diphosgene, 99%. Catalog Numbers: XXXXXX, XXXXXXX. Synonyms: Trichloromethyl chloroformate; carbonochloridic acid, trichloromethyl ester. Molecular formula: $C_2C_{l4}O_2$. Molecular weight: 197.83 g. Company Identification: Company Name: XXXX Company. Company Address 1: Company Address 2: For information in North America, call: 1-XXX-XXX-XXXX. For emergencies in the U.S., call: 1-XXX-XXX-XXXX.
Section 2: Composition/Information on Ingredients	
Chemical names, Chemical Abstract Service (CAS) numbers and percentages of chemicals that comprise the product (if the material is a compound) are found in this section. For a carcinogen identified by the National Toxicology Program (NTP), International Agency for Research on Cancer (IARC), or the U.S. Occupational Safety and Health Administration (OSHA), its presence is noted here, when ≥0.1%. If the makeup of the chemical is a trade secret, its composition may not be listed, but hazards and safety precautions are found here.	**Chemical Name:** Diphosgene. **CAS#:** 503-38-8. **Percent:** 99%.

(continued)

Material Safety Data Sheet (Continued)

Section 3: Hazards Identification

The material's appearance, odor, and health, physical, and environmental hazards that may be of concern for emergency response personnel are found here. Understanding the consequences of exposure to this substance is important.

Appearance: Clear, colorless liquid.

Danger! Corrosive. Causes eye and skin burns. May cause severe respiratory tract irritation with possible burns. May cause severe digestive tract irritation with possible burns. Lacrimator (substance that increases the flow of tears). May cause sensitization by inhalation and skin contact.

Reacts violently and/or explosively with water, steam, or moisture.

Target Organs: None.

Potential Health Effects:

Eye: Causes eye burns. Lacrimator.

Skin: Causes skin burns. May cause skin sensitization, an allergic reaction that becomes evident upon reexposure to this material.

Ingestion: May cause severe and permanent damage to the digestive tract. Causes gastrointestinal tract burns. Direct aspiration into the lungs may cause chemical pneumonitis, pulmonary edema, and hemorrhaging.

Inhalation: Causes chemical burns to the respiratory tract. Inhalation may produce coughing, nausea, and pulmonary edema. In rare instances, exposure may cause sensitization, resulting in inflammation of the mucous membranes and in eczematous eruptions.

Chronic: Repeated or prolonged exposure may cause allergic reactions in sensitive individuals.

Section 4: First-Aid Measures (or Health Hazards)

Information regarding acute and chronic health effects and what should and should not be done when a person is exposed to the chemical is found in this section. Since chemicals can enter the human body through different routes, procedures dealing with each potential route of exposure are also included here.

Eyes: Get medical aid immediately. Do NOT allow victim to rub or keep eyes closed. Extensive irrigation with water is required (at least 30 minutes).

Skin: Get medical aid immediately. Immediately flush skin with plenty of water for at least 15 minutes while removing contaminated clothing and shoes. Wash clothing before reuse. Destroy contaminated shoes.

Ingestion: Do NOT induce vomiting. If victim is conscious and alert, give 2 to 4 cupfuls of milk or water. Never give anything by mouth to an unconscious person. Get medical aid immediately.

(continued)

Material Safety Data Sheet (Continued)

Section 4: First-Aid Measures (or Health Hazards) (Continued)

Inhalation: Get medical aid immediately. Remove from exposure and move to fresh air immediately. If not breathing, give artificial respiration. If breathing is difficult, give oxygen. Do NOT use mouth-to-mouth resuscitation.

Notes to Physician: Treat symptomatically and supportively.

Section 5: Fire-Fighting Measures (or Physical Hazards)

This section describes potential flammability and explosion hazards. Recommended extinguishing media to be used, procedures in handling these hazards, and storage considerations are included here. It may also contain information on chemical incompatibility. This section contains important characteristics of the chemical, such as extinguishing media, flash point, flammable limits (explosive limits), ignition temperature, auto-ignition temperature, hazardous decomposition products, fire-fighting equipment, fire-fighting methods, Fire Protection Association (FPA) codes, chemical incompatibilities, chemical instabilities, and hazardous polymerization.

General Information: Containers can build up pressure if exposed to heat and/or fire. As in any fire, wear a self-contained breathing apparatus in pressure-demand, Mine Safety and Health Administration (MSHA)/NIOSH (approved or equivalent), and full protective gear. During a fire, irritating and highly toxic gases may be generated by thermal decomposition or combustion. Use water spray to keep fire-exposed containers cool. Vapors may be heavier than air. They can spread along the ground and collect in low or confined areas. May ignite or explode on contact with steam or moist air. Substance may react with water, and may release corrosive and/or toxic gases.

Extinguishing Media: Use dry sand or earth to smother fire. Do NOT get water inside containers. For large fires, use water spray, fog, or regular foam. Contact professional fire fighters immediately. Cool containers with flooding quantities of water until well after fire is out. For small fires, use dry chemical or carbon dioxide.

Flash Point: Not available.

Auto-ignition Temperature: Not available.

Explosion Limits: Not available.

Lower: Not available.

Upper: Not available.

National Fire Protection Association (NFPA) Rating (estimated): Health: 3; Flammability: 1; Instability: 1; Special Hazard: W.

Section 6: Accidental Release Measures

Information for preventing or minimizing adverse effects in the event of a spill or leak and methods of cleanup and disposal are presented in this section.

General Information: Use proper personal protective equipment as indicated in Section 8.

(continued)

Material Safety Data Sheet (Continued)

Section 6: Accidental Release Measures (Continued)

Spills/Leaks: Absorb spill with inert material (e.g., vermiculite, sand, or earth), and then place in suitable container. Avoid runoff into storm sewers and ditches that lead to waterways. Clean up spills immediately, observing precautions in Section 8. Isolate area and deny entry. Provide ventilation. Do not expose spill to water. Do not get water inside containers. A vapor-suppressing foam may be used.

Section 7: Handling and Storage

This section provides guidelines for minimizing potential hazards from storage and handling the chemical. Conditions for storage and safe handling and conditions that should be avoided are listed.

Handling: Wash thoroughly after handling. Remove contaminated clothing and wash before reuse. Use only in a well-ventilated area. Do not allow water to get into the container because of violent reaction. Avoid contact with skin and eyes. Keep container tightly closed. Avoid ingestion and inhalation. Do not allow contact with water. Discard contaminated shoes. Keep from contact with moist air and steam.

Storage: Keep container closed when not in use. Store in a cool, dry, well-ventilated area away from incompatible substances. Keep away from water.

Section 8: Exposure Controls, Personal Protection

Engineering control needed when handling the material and personal protective equipment (e.g., respirators, gloves, and goggles) to be used are discussed. Exposure and work practices guidelines are included in this section. Specific steps and precautions to take to safeguard human health, such as proper ventilation requirements, are included here.

Engineering Controls: Use explosion-proof ventilation equipment. Facilities storing or utilizing this material should be equipped with an eyewash facility and a safety shower. Use adequate ventilation to keep airborne concentrations low.

Exposure Limits:

American Conference of Industrial Hygienists (**ACGIH**): None listed.

NIOSH: None listed.

OSHA Final PELs: None listed.

OSHA Vacated PELs: No OSHA vacated PELs are listed for this chemical.

Personal Protective Equipment:

Eyes: Wear appropriate protective eyeglasses or chemical safety goggles as described by OSHA's eye and face protection regulations in 29 CFR 1910.133 or European Standard EN166.

Skin: Wear appropriate protective gloves to prevent skin exposure.

(continued)

Material Safety Data Sheet (Continued)

Section 8: Exposure Controls, Personal Protection (Continued)

Clothing: Wear appropriate protective clothing to minimize contact with skin.
Respirators: Wear a NIOSH/MSHA or European Standard EN 149 approved full-face airline respirator in the positive pressure mode with emergency escape provisions.

Section 9: Physical and Chemical Properties

This section includes physical and chemical properties of the chemical. These properties may include many items such as physical state (liquid, solid, gas), appearance, odor, physical state, pH, vapor pressure and density, evaporation rate, viscosity, boiling point, melting and freezing point, decomposition temperature, solubility, and specific gravity, among others. Other useful properties may also be found here.

Physical State: Liquid.
Appearance: Clear, colorless.
Odor: Phosgene odor.
pH: Not available.
Vapor Pressure: 13.7 mbar at 20°C.
Vapor Density: 6.8.
Evaporation Rate: Not available.
Viscosity: Not available.
Boiling Point: 128°C at 760.00 mmHg.
Freezing/Melting Point: –57°C.
Decomposition Temperature: Not available.
Solubility: May decompose.
Specific Gravity/Density: 1.6400g/cm^3.

Section 10: Stability and Reactivity

Conditions under which the chemical is stable or unstable or a potentially hazardous reaction may occur, such as storage and handling temperature, reactions with other chemicals, production of heat, or other hazardous conditions are given in this section.

Chemical Stability: Stable under normal temperatures and pressures. Combines vigorously or explosively with water.
Conditions to Avoid: Incompatible materials, excess heat, exposure to moist air or water, oxidizers.
Incompatibilities with Other Materials: Metals, strong oxidizing agents, ammonia, isopropyl alcohol.
Hazardous Decomposition Products: Hydrogen chloride, phosgene, irritating and toxic fumes and gases, hydrogen gas.
Hazardous Polymerization: Has not been reported.

Section 11: Toxicological Information

This section includes information on toxicity of the chemical, and its potential for causing cancer, and other chronic health problems. Acute, subchronic, and chronic exposure data can also be found here.

Registry of Toxic Effects of Chemical Substances (RTECS)#: LQ7350000
CAS#: 503-38-8: LD_{50}/LC_{50}: Not available.
Carcinogenicity:
CAS# 503-38-8: Not listed by ACGIH, IARC, NIOSH, NTP, or OSHA.
Epidemiology: No information available.
Teratogenicity: No information available.
Reproductive Effects: No information available.
Neurotoxicity: No information available.
Mutagenicity: No information available.
Other Studies: No data available.

(continued)

Material Safety Data Sheet (Continued)

Section 12: Ecological Information

Impacts to the environment that may occur if the material leaks to the environment are discussed here.

Ecotoxicity: No data available. No information available.
Environmental: Based on its low melting and boiling points, this chemical is expected to rapidly volatilize into the atmosphere.
Physical: No information available.
Other: No information available.

Section 13: Disposal Considerations

This section includes guidance for environmental and other technical personnel responsible for waste management of this chemical.

Chemical waste generators must determine whether a discarded chemical is classified as a hazardous waste. U.S. EPA guidelines for classification determination are listed in 40 CFR Parts 261.3. Additionally, waste generators must consult state and local hazardous waste regulations to ensure complete and accurate classification.
RCRA P-Series: None listed.
RCRA U-Series: None listed.

Section 14: Transport Information

This section gives information regarding U.S. Department of Transportation (DOT) and other classifications.

U.S. DOT.
International Air Transport Association (IATA).
European Agreements Governing the Regulations Concerning the International Transport of Dangerous Goods by Raid (RID) or Road (ADR) (RID/ADR).
International Maritime Organization (IMO).
Canada, Transport Dangerous Goods (TDG).
Shipping Name: TOXIC LIQUIDS, CORROSIVE, ORGANIC, Not otherwise specified N.O.S.:
No information available.
Hazard Class: 6.1.
United Nations (UN) Number: UN2927.
Packing Group: I

Section 15: Regulatory Information

Information regarding the regulatory status of the material: OSHA regulations, EPA regulations, regulations by other federal regulatory agencies, and state agencies should be found in this section.

U.S. FEDERAL REGULATIONS
Toxic Substances Control Act (TSCA). CAS# 503-38-8 is listed on the TSCA inventory.
Health & Safety Reporting List: None of the chemicals in this product are on the Health & Safety Reporting List.

(continued)

Material Safety Data Sheet (Continued)

Section 15: Regulatory Information (Continued)

Chemical Test Rules: None of the chemicals in this product are under a Chemical Test Rule.

Section 12b: None of the chemicals are listed under TSCA Section 12b.

TSCA Significant New Use Rule (SNUR): None of the chemicals in this material have a SNUR under TSCA.

Comprehensive Environmental Response Compensation & Liability Act (CERCLA) Hazardous Substances and corresponding Reportable Quantity (RQ): None of the chemicals in this material have an RQ.

Superfund Amendments and Reauthorization Act (SARA) Section 302 Extremely Hazardous Substances: None of the chemicals in this product have a threshold planning quantities (TPQ).

Section 313: None of the chemicals in this product are reportable under Section 313.

Clean Air Act (CWA): This material does not contain any hazardous air pollutants. This material does not contain any Class 1 ozone depletors. This material does not contain any Class 2 ozone depletors.

None of the chemicals in this product are listed as Hazardous Substances under the CWA. None of the chemicals in this product are listed as Priority Pollutants under the CWA. None of the chemicals in this product are listed as Toxic Pollutants under the CWA.

Occupational Safety and Health Administration (OSHA): None of the chemicals in this product are considered highly hazardous by OSHA.

STATE REGULATIONS

CAS# 503-38-8 is not present on state lists from CA, PA, MN, MA, FL, or NJ. None of the chemicals in this product are listed on CA state list.

European/International Regulations: European Labeling in Accordance with EC Directives.

Hazard Symbols: T+ C.

Risk Phrases: R 26/27/28, Very toxic by inhalation, in contact with skin and if swallowed. R 34, Causes burns.

Safety Phrases: S 36/39, Wear suitable protective clothing and eye/face protection. S 45, In case of accident or if you feel unwell, seek medical advice immediately (show the label where possible).

(continued)

Material Safety Data Sheet (Continued)

Section 15: Regulatory Information (Continued)

WGK (Water Danger/Protection)
CAS# 503-38-8: No information available.
Canada, Domestic Substance List and the Nondomestic Substance List (DSL/NDSL): CAS# 503-38-8 is listed on Canada's NDSL list.
Canada, Workplace Hazardous Materials Information System (WHMIS): This product does not have a WHMIS classification.
Canadian Ingredient Disclosure List.
Exposure Limits.

Section 16: Other

This last section may include pertinent information that does not fit in above sections. Information such as label information, hazard ratings, revision dates, and references to other related information should be provided.

MSDS Creation Date: 10/20/1998.
Revision #2 Date: 3/18/2003.
The information above is believed to be accurate and represents the best information currently available to us. However, we make no warranty of merchantability or any other warranty, express or implied, with respect to such information, and we assume no liability resulting from its use. Users should make their own investigations to determine the suitability of the information for their particular purposes. In no event shall the company be liable for any claims, losses, or damages of any third party or for lost profits or any special, indirect, incidental, consequential or exemplary damages, howsoever arising, even if the company has been advised of the possibility of such damages.

APPENDIX B

Standing Operating Procedure

Building #:________

Room(s) #:________

(Check all that apply)

____Amendment

____Chemical Agent

____Biological/Toxin

____Radiation

____Lasers

____Industrial

____Pyrotechnics

____Other (specify:__________)

ABCD COMPANY

Title: Operation of Toxic Vapor Generator

SOP#: XXX-XXX

Submitted by:

____
Team Leader

Environmental Quality Office: ____________________________

Risk Reduction Office: ____________________________

Approved by

____
Director

APPROVAL DATE: ____________________________

SOP TITLE

Operation of Toxic Vapor Generator

I have read this SOP.

Signature	Date

Title: Operation of Toxic Vapor Generator

I. **Statement of Work** (Describe the intent of the SOP. An example follows.)

This SOP involves the operation of a multipurpose chemical vapor generation system for generating controlled toxic vapor concentrations over a wide range, in an approved hood or agent test chamber. This vapor generation system is primarily used for testing detection and alarm systems.

II. **Additional Nonstandard Responsibilities** (List responsibilities of supervisors, operators, and all organizations involved in this operation)

III. **Materials** (List necessary supplies, equipment, and substances to be used to support the operation. An example follows.)

A. General laboratory chemicals
B. Toxic chemical surety material (List surety compounds to be used.)
C. Industrial chemicals (List all other TICs as needed.)
D. Experimental compounds (List all experimental compounds considered.)

IV. **Hazards Involved**

A. General hazards involved
B. Specific hazards: Refer to appropriate MSDS for chemicals.
C. Equipment hazards: Refer to operator's manual for equipment hazards.

VI. **Safety Requirements** (List safety requirements and/or related documents.)

VII. **Procedures** (Explain the process. An example follows.)

The humidified agent vapor generator is used to provide a source of controlled humidified toxic chemical agent vapor for use in testing agent detection and warning devices. The agent vapor provided at the "sample port" may be delivered directly to the device being tested or to an optional glass cup or "test chamber" in which the device will be challenged. The "agent reservoir" can be either an agent bubbler or a "delta tube" that is loaded with a small amount of the chemical. An airstream passing over the deposited agent will carry the chemical vapor to the mixing manifold for dilution with a relative-humidity conditioned airstream to achieve the desired concentration.

Preagent Operation Inspection and Test (Describe procedure to ensure integrity of the system and explain. An example follows.)

Set up equipment; examine all parts of the generator; label fittings of the agent reservoir and all generator connections; check for flow restrictions, loose connections, and proper flow connections throughout the system. List steps to complete the setup before adding surety material to the system (refer to Checklist A at the end of this SOP).

List Operational Procedure (List step-by-step and describe each step. Examples follow.)

A. *Secure vapor source from storage hood:* Retrieve and transport the chemical agent from the "storage hood" to a working hood. The agent must be doubly contained during the transfer. Be sure to have a mask and a buddy available. Wear (TAP) or "approved" gloves for handling the agent container.

B. *Check agent reservoir and agent bottle for external liquid contamination on outside of agent container* (example follows): Check agent reservoir and agent bottle for external liquid agent contamination on outside of agent container using the M8 paper. Wrap the top of the agent storage bottle with a piece of parafilm and place in a metal screw- or friction-top can containing vermiculite before returning the surety material to storage refrigerator. Do not touch the can with potentially contaminated gloves. Submerge all waste in a decontamination bath, wet thoroughly with decontamination solution, and dispose of in accordance to regulations.

C. *Operation of vapor generator:* (Provide details of operation. Example follows.) In general, the operation would be performed by qualified personnel. Operators will read, sign, and observe the safety procedures listed in the SOP.

D. *Sampling of agent vapors:* (List steps used for testing and sample collection and analytical methodology. An example for solid sorbent sampling of chemical vapor follows.)

   a. A solid sorbent tube packed with a suitable sorbent is connected to a filtered suction line.

   Flow rate through the sorbent tube is accurately measured.

   b. The sorbent tube is then inserted to the sample port to draw the sample for a length of time determined by the expected vapor concentration.
   c. After the sample period, the sorbent tube is heat desorbed into a GC-type of instrument for analysis.

E. *Decontamination:* Surety materials to be decontaminated will be collected in the appropriate decon solution following the established decontamination procedures.

F. *Waste Treatment and Disposal* (example follows):

   a. Solid wastes are removed from the decontamination bath, air dried, and collected in double-bag–lined toxic waste cans before capping and sealing with masking tape. The cans will be placed in fiberboard drums double-lined with plastic bags. The drums will then be transferred to the decon/detox facility for incineration or proper disposal. Waste containing arsenic and Whetlerite carbon will be containerized separately and will not be incinerated. A hazardous waste contractor shall dispose of this waste. The spent decontamination solution, or liquid waste, will be combined with other spent decontamination solutions for proper disposal.

b. Spent liquid decontamination solutions (bleach and/or sodium hydroxide) are placed in a jar, double-bagged and overpacked with vermiculite in a toxic waste can. The can is disposed of as per appropriate regulations.

G. *Emergency shutdown procedures:* (List procedures to follow in case of power failure, fire, or other emergency incidents. An example follows.) In the event of an agent spill outside engineering controls, follow emergency procedures in the installation's emergency evacuation plan.

VIII. Emergency First-Aid/Self-Aid Procedures

See *Related Document ABC (example)* for first-aid/self-aid procedures to be used in the event of emergency conditions arising during operations described in this SOP.

IX. Decontamination and Disposal of Toxic Chemical Agents

A. *Decontamination* (List procedure or related documents for decontamination procedure.)
B. *Disposal* (List procedure or related documents for decontamination procedure.)

X. Site-Specific Contingency Plan for Spill Response

A. In case of a toxic or hazardous chemical spill resulting from fire, explosion, or other cause, the person discovering the spill will take whatever measures are immediately available to stop the discharge and contain the spilled material. This should be done only if doing so does not endanger the person's own health or safety or that of other employees in the area. The person will then immediately report the spill to:

1. Installation-wide emergency number, XXX-XXXX or 911.
2. Risk Reduction Office, XXX-XXX-XXXX.
3. Environmental Quality Office, XXX-XXX-XXXX.

B. Procedures outlined in Documents D, E, and F (examples), which are posted in all laboratory rooms, will then be followed. Action will be coordinated with the Risk Reduction Office and Environmental Quality Office.

CHECKLIST A

Hazardous Operation (Surety)

Preoperational Checklist

Ensure That Is Present in the Room Prior to Start of Operations

General

1. SOP contents applicable? __________
2. SOP current? __________
3. CSM use permit available and posted? __________
4. Posted at entrance? __________
5. Inventory record current? __________
6. First-aid kit available and updated? __________
7. Eyewash and shower tested? __________
8. Equipment properly marked? __________
9. Equipment in good repair? __________
10. Buddy present? __________
11. Supervisor notified? __________

Ventilation

1. Ventilation system approved? Hood # ______ Avg Velocity ______
2. Alarm activated? __________
3. Sash at approved level? __________
4. Hood face velocity checked? __________

SOP Operation Preliminary

1. Proper decon material available? ________
2. Toxic and hazardous waste container available? ________
3. Protective mask and clothing available? ________
4. Approved gloves available? ________
5. Eye protection used? ________
6. Lab coat worn? ________
7. Detection equipment available? ________

Proceed to Operation:

1. Visual examination of all parts of generator? ________
2. Generator ports labeled properly? ________
3. Necessary humidifier and dehumidifiers charged? ________
4. Check all lines for leaks or restrictions? ________
5. Test generator for proper flows as desired? ________
6. Turn off generator before connecting agent reservoir? ________
7. Agent reservoir attached properly and securely? ________
8. Double-check above steps? ________

Operators: 1. __________ 2. __________ Date: __________

CHECKLIST B

Hazardous Operation (Surety)

Postoperational Checklist

SOP # ______________

DATE: ______________

1. Generator turned off? __________
2. Agent reservoir properly capped or contained? __________
3. Gloves and other waste properly disposed? Registered? __________
4. Area cleaned up? __________
5. CSM material properly secured and accounted for? __________
6. Observed symptoms of possible exposure in buddy? __________
7. Washed hands with soap and water? __________
8. Secure hoods and room? __________

Operator 1. ____________________

Operator 2. ____________________

Date: ________________________

Time: ________________________

APPENDIX C

Index of Chemical Agent Detectors and Supplies and Manufacturers

Chemical Agent Detectors and Supplies	Manufacturers/Distributors
ACADA GID-3 Automatic Chemical Agent Detection Alarm	Smiths Detection
AP2C Vapor Detector	Proengin Inc.
APD-2000 Handheld Detector	Smiths Detection
CAM/ICAM/CAM2 Chemical Agent Monitor	Smiths Detection
Chemical Agent Liquid Detectors, 3 Way	Tradeways Ltd.
Chemical Agent Water Test Kit M272	Tradeways Ltd.
Chemical Warfare Agent Detector	Environics USA
ChemPro 100 Detector	Environics USA
Cyranose™ 320 Portable Handheld Device	Cyranose Sciences Inc.
Dynatherm-GC-FPD/MS	Agilent Technologies
HAWK Long-Range Chemical Detector	Bruker Daltonics Inc.
HAZMATCAD and HAZMATCAD/Plus Detectors	Microsensor Systems
HazMatIDTM/TravelIR-HCITM Chemical Identifier	SensIR Technologies
HyFed	Tracor
IMS 2000 Handheld Chemical Warfare Agent Detector	Bruker Daltonics Inc.
Individual Chemical Agent Detector (ICAD)	Smiths Detection
Innova Photoacoustic Infrared Spectroscopy	Innova (Denmark)
Innova Multigas Monitors, models 1312 and 1314	Innova
IPDS Improved Point Detection System	Tradeways Ltd.
JCAD Joint Chemical Agent Detector	BAE Systems

(continued)

Chemical Agent Detectors and Supplies	Manufacturers/Distributors
JSLSCAD Joint Service Lightweight Standoff Chemical Agent Detector	General Dynamics
LCD Lightweight Chemical Agent Detector	Smiths Detection
M18	Tradeways Ltd.
M18A1	Tradeways Ltd.
M18A3 and M256A1 Chemical Agent Detector Kits	Tradeways Ltd.
M256A	Tradeways Ltd.
Chemical Agent Water Test Kit M272	Tradeways Ltd.
M43	Tradeways Ltd.
M43A1	Tradeways Ltd.
M43 Upgrade Chemical Agent Detector	General Dynamics
M8	Tradeways Ltd.
M9	Tradeways Ltd.
M90	Environics USA
M-90A1-C Chemical Warfare Agent Detector	Environics USA
Micro-Chem Lab	Sandia National Laboratories
MicroFID	PerkinElmer
MiniCAD	Microsensor Systems
MINICAMS Model FM-3001	CMS Field Products
MIRAN SapphIRe Portable Ambient Air Monitor (Models 1, 5, 30, 30E, 100, and 100E)	Foxboro Co.
RAID-M Handheld Chemical Agent Detector	Bruker Daltonics Inc.
RAID-S Stationary Rapid Alarm and Identification Device	Bruker Daltonics Inc.
Rapid Air Pollution Infrared Detector (RAPID)	Bruker Daltonics Inc.
Remote Sensing Chemical Agent Alarm, 21	Tradeways Ltd.
S4PE	Proengin Inc.
Sabre-2000 Vapor Detector	Smiths Detection
CW Sentry Plus	Microsensor Systems
Simultaneous Test Sets from Draeger Safety	AFC International, Inc.
Swagelok	Swagelok Co.
Teflon	DuPont
TIMs Detector Toxic Industrial Materials Detector	Proengin Inc.
Transport IR-100 Spectrophotometer	Thermo Nicolet Corp.
Transport Kit	Thermo Nicolet Corp.
Tygon	Saint-Gobain Ceramics & Plastics, Inc.
UC-AP2C	Proengin Inc.
Travel*IR*	SensIR Corp.
DuraDisk	SensIR Corp.
TVA-1000B Toxic Vapor Analyzer	Foxboro Co.
Passport II	Mine Safety
ppbRae	Rae Systems
Hnu Model PI-101 Trace Gas Analyzer	HNU System, Inc.

Glossary and Abbreviations

AC Hydrogen cyanide.

ACADA Automatic Chemical Agent Detection Alarm, in use by U.S. military.

Acetylcholine A chemical neurotransmitter produced by nerve cells predominantly outside the central nervous system. It is a chemical "messenger," stimulating the heart, skeletal muscles, and numerous secretory glands.

ACT Applied Chemistry Team, established by U.S. Army Soldier and Biological Chemical Command at Aberdeen Proving Ground, Edgewood, Maryland. Formerly known as Design Evaluation Laboratory.

AEGLs Acute exposure guideline levels.

AEL Airborne exposure concentration limit; causes no long-term effects after exposure of up to 8 hr.

Aerosol A liquid or solid composed of finely suspended particles in a gaseous medium.

Agent Chemical warfare compound or surety material.

amu Atomic mass unit.

APG Aberdeen Proving Ground, Edgewood, Maryland.

atm Atmospheric pressure at sea level, equal to 760 mmHg.

Blister agent A chemical agent that injures the eyes and lungs and blisters the skin.

Carcinogen Substance that has the potential to cause cancer in humans.

Carcinogenicity The potential for development of cancer in humans.

CAS Chemical Abstract Service.

CG Phosgene.

Chemical contamination Presence of chemical agent in environment (air, water, soil, food, clothing, etc.) or persons.

Chemical warfare All aspects of military operations involving the use of lethal agents.

Chemical warfare agent (CWA) Chemical compound used with intent to kill, injure, or incapacitate persons in military operations.

CK Cyanogen chloride.

CWAs Chemical warfare agents.

DA Diphenylchlorarsine.

Decontamination Process of reducing the amount of chemical agent on objects, persons, and/or areas.

Decontamination level Designates degree of decontamination expressed as 1X, 3X (XXX), 5X (XXXXX), 0.

Detection Determine presence of a chemical agent.

DoD U.S. Department of Defense.

DPG Dugway Proving Ground, under the U.S. Army Test and Evaluation Command, located in Utah.

DPP Domestic Preparedness Program.

DTS Detector testing system.

EC Electrochemical.

ECBC U.S. Army Edgewood Chemical and Biological Center, Maryland.

EPA U.S. Environmental Protection Agency.

Exclusion area Immediate area that has access to chemical agents.

Exposure Exposed to chemical agent.

FID Flame ionization detector.

FPD Flame photometric detector.

FT-IR Fourier transform infrared spectroscopy.

GA Code designation for CAS 77-81-6, a nerve agent.

G-agents Collective name for GA, GB, GD, and GF.

GB Code designation for CAS 107-44-8, a nerve agent.

GC Gas chromatography.

GD Code designation for CAS 96-64-0, a refractory nerve agent.

GF Code designation for CAS 329-99-7, a nerve agent.

h Planck's constant.

H Levinstein mustard with about 30% sulfur impurity, a blister agent.

HD Distilled Levinstein mustard CAS 505-60-2, a blister agent.

HN-1 Nitrogen mustard, 2,2-Dichlorotriethylamine, a blister agent.

HN-2 Nitrogen mustard, 2, 2′-Dichloro-N-methyldiethylamine, a blister agent.

HN-3 Nitrogen mustard, 2,2′,2″-trichlorotriethylamine, a blister agent.

HT Mixture of 60% HD and 40% T (Bis-[2(2-chloroethylthio)ethyl] ether, with CAS63918-89-8; a blister agent.

IC_{50} Concentration of chemical agent that would incapacitate 50% of exposed population.

ICt_{50} Dosage that incapacitates 50% of the exposed population at concentration multiplied by the exposure time.

IDLH Immediate danger to life and health.

IDLH concentration Airborne concentration that causes no long-term effect on exposure of up to 30 min.

IMS Ion mobility spectroscopy.

Incapacitating agent Agent that temporarily disables a person. Casualties can recover without medical attention.

Industrial chemical Chemical developed for industrial uses, rather than specifically to cause human casualties.

IR Infrared.

JCAD Joint Chemical Agent Detector, a multimission chemical agent point detection system, currently in development for the U.S. military.

JSOR Joint Services Operational Requirements.

L Lewisite; code designation for CAS541-25-3, a blister agent.

LCt_{50} Dosage received as the product of concentration multiplied by the exposure time that is lethal to 50% of exposed population. See also LD_{50}.

LD_{50} Lethal dosage of chemical agent for 50% of exposed population.

MDL Minimum detectable level.

mg/m^3 Milligrams per cubic meter.

mg-min/m^3 Expression for dosage: milligrams multiplied by exposure minutes per cubic meter.

Monitoring Continuous or periodic determination of presence of chemical agent.

MSDS Material Safety Data Sheet.

MTBFAs Mean time between false alarms.

MW Molecular weight.

Nerve agent A lethal agent that interferes with the ability of muscles to relax.

NIOSH National Institute for Occupational Safety and Health.

NIST National Institute of Standards and Technologies.

Nitrogen mustards Collective name for HN-1, HN-2, and HN-3.

OLES Office of Law Enforcement Standards, NIST.

ORM Operational risk management.

OSHA Occupational Safety and Health Administration.

PEL Permissible exposure limit; TWA concentration that must not be exceeded during an 8-hr workday in a 40-hr week.

Persistency An expression of duration of effectiveness for a chemical agent. May denote class of chemical agent.

PFPD Pulsed flame photmetric detector.

PID Photo ionization detector.

PMT Photomultiplier tube.

ppb Parts per billion.

ppm Parts per million.

Refractory compounds Chemicals that accumulate (i.e., are not released/removed from tissue) in the human body.

REL Recommended exposure limit; TWA concentration for up to a 10-hr workday during a 40-hr workweek that should not be exceeded.

RH Relative humidity.

SA Arsine, Arsenic trichloride, a blood agent, CAS 7784-42-1.

SAW Surface acoustic wave.

SBCCOM U.S. Army Soldier Biological and Chemical Command, Aberdeen Proving Ground, Edgewood, Maryland.

SOP Standing operating procedure.

STEL Short-term exposure limit designed for a 15-min TWA exposure that should not be exceeded at any time during a workday.

TIC Toxic industrial compound or chemical.

TIM Toxic industrial material.

Toxicity The property of a substance that injures an organism's physiological mechanism(s) to cause incapacitation or death.

TWA Time-weighted average.

USACHPPM U.S. Army Center for Health Promotion and Preventive Medicine, Aberdeen Proving Ground, Edgewood, Maryland.

V-agents Collective name for VX and Vx, a nerve agent.

Vapor pressure (VP) Pressure exerted by a vapor in equilibrium between vapor and its liquid and solid state.

VD Relative vapor density; vapor and air density ratio.

Volatility (V) Weight of vapor in a unit volume at equilibrium at a specific temperature.

VX Code designation for CAS 50782-69-9, a nerve agent.

WMD Weapon of mass destruction.

Bibliography

Amirav, A., Pulsed flame detector method and apparatus, U.S. patent no. 5153673, Israel patent no. 95617, European patent no. 0475250, Japan patent no. 2759854.

Asbury, G.R., et al., Separation and identification of some chemical warfare degradation products using electrospray high resolution ion-mobility spectrometry with mass selected detection. *Anal. Chim. Acta*, 404 no. 2, 273–283, 2000.

Buchanan, J.H., Buettner, L.C., Butrow, A.B., Vapor pressure of VX, U.S. Army Edgewood Chemical and Biological Center, Maryland, ECBC-TR-068, 1999.

Causey, M.T. and Buettner, L.C., Pressure swing adsorption filtration proto-type scale. Study IV: 2 Hexanol and isopropyl/methyl phosphonofluoridate (GB) pulse feed, ERDEC-TR-345, U.S. Army Edgewood Research, Development and Engineering Center (ERDEC), Aberdeen Proving Ground, Maryland, 1996.

Eiceman G.A. and Karpas Z., *Ion mobility spectrometry*, CRC Press, Boca Raton, FL, 1994.

Ellison, D.H., *Handbook of chemical and biological warfare agents*, CRC Press, Boca Raton, 2000.

Ewing, R.G. and Miller, C.J., Detection of volatile vapours emitted from explosives with a handheld ion-mobility spectrometer, *Field Anal. Chem. Technol.*, 5, 215–221, 2001.

Frishman, G. and Amirav, A., Fast GC-PFPD System for field analysis of chemical warfare agents, *Field Anal. Chem. Technol.*, 4, 170–194, 2000.

Handley, A.J. and Adlard, E.R., *Gas chromatographic techniques and applications*, CRC Press, Boca Raton, FL, 2001.

Hibbert, D.B., Detecting drugs at customs, *Chem. Austral.*, 66, 8–9, 1999.

Hooijschuur, E.W.J., Kientz, C.E., Brinkman, U.A.T., Determination of alkylphosphonic acids by microcolumn liquid chromatography with gradient elution coupled online with flame photometric detection, *J. Chromatogr. A*, 907, 165–172, 2001.

Hooijschuur, E.W.J., Kientz, C.E., Brinkman, U.A.T., Application of microcolumn liquid chromatography and capillary electrophoresis with flame photometric detection for the screening of degradation products of chemical warfare agents in water and soil, *J. Chromatogr. A*, 928, 187–199, 2001.

Isson, R.M., et al., Improvement of instrumental methods, DPG-FR-88-711, U.S. Army Test and Evaluation Command, Dugway Proving Ground, Utah, 1988.

Lewis, W.S., Sander, J.R., Houle, M.J., The use of permeation tubes as primary standards in the calibration of flame photometric detectors at low concentrations, TR-DPG L-135, U.S. Army Test and Evaluation Command, Dugway Proving Ground, Utah, March 1979.

Jing, H. and Amirav, A., Pulsed flame photometric detector — a step forward towards universal heteroatom selective detection, *J. Chromatog. A*, 805, 177–215, 1998.

Karwacki, C.K., Buettner, L.C., Mahle, J.J., Adsorption equilibria of isopropyl methylphosphonofluoridate (GB) on activated carbon at ultra-low relative pressures, CJK, DET, JJM, JHB, and LCB, *Langmuir*, 15, 6343–6345, 1999.

Keller, T., et al., Detection of designer drugs in human hair by ion mobility spectrometry (IMS). *Forensic Sci. Int.*, 94, 55–63, 1998.

Klinghoffer, A.A. and Rossin, J.A., Catalytic oxidation of a CK model compound over a monolithic oxidation catalyst, ERDEC-CR-041, U.S. Army Edgewood Research, Development and Engineering Center (ERDEC), Aberdeen Proving Ground, Maryland, May 1993.

Longworth, T.L., Cajigas, J.C., Barnhouse, J.L., Testing of commercially available detectors against chemical warfare agents: summary report, Soldier and Biological Chemical Command, Aberdeen Proving Ground, Maryland, February 1999.

Longworth, T.L., Barnhouse, J.L., Ong, K.Y., Domestic Preparedness Program: testing of photovac microfid handheld flame ionization detectors against chemical warfare agents — summary report, Engineering Directorate, Soldier and Biological Chemical Command, Aberdeen Proving Ground, Maryland, October 1999.

Longworth, T.L., Baranoski, J.M., Ong, K.Y., Domestic Preparedness Program: testing of MIRAN SapphIRe Portable Ambient Air Analyzers against chemical warfare agents summary report, Research and Technology Directorate, Soldier and Biological Chemical Command, Aberdeen Proving Ground, Maryland, July 2000.

Longworth, T.L., Baranoski, J.M., Ong, K.Y., Domestic Preparedness Program evaluation of the Agilent GC–FPD/MSD (gas chromatograph–flame photometric detector/mass selective detector) system against chemical warfare agents summary report, Research and Technology Directorate, Soldier and Biological Chemical Command, Aberdeen Proving Ground, Maryland, October 2002.

Longworth, T.L. and Ong, K.Y., Domestic Preparedness Program: testing of detectors against chemical warfare agents — summary report, UC AP2C \Portable Chemical Contamination Control Monitor Collective Unit, Research Directorate, Soldier and Biological Chemical Command, Aberdeen Proving Ground, Maryland, May 2001.

Longworth, T.L. and Ong, K.Y., Domestic Preparedness Program: testing of the CAM-Chemical Agent Monitor (type L) against chemical warfare agents summary report, Research Directorate, Soldier and Biological Chemical Command, Aberdeen Proving Ground, Maryland, August 2001.

Longworth, T.L., Ong, K.Y., Baranoski, J.M., Domestic Preparedness Program: testing of M90-D1-C Chemical Warfare Agent Detector against chemical warfare agents summary report, Research and Technology Directorate, Soldier and Biological Chemical Command, Aberdeen Proving Ground, Maryland, December 2000.

Longworth, T.L. and Ong, K.Y., Domestic Preparedness Program: testing of Sabre 2000 Handheld Trace And Vapor Detector Against Chemical Warfare Agents Summary Report, Research Directorate, Soldier and Biological Chemical Command, Aberdeen Proving Ground, Maryland, August 2001.

Longworth, T.L. and Ong, K.Y., Domestic Preparedness Program: testing of RAE Systems PPBRAE Volatile Organic Compound (VOC) Monitor Photo-Ionization Detector (PID) against chemical warfare agents summary report, Research Directorate, Soldier and Biological Chemical Command, Aberdeen Proving Ground, Maryland, September 2001.

Longworth, T.L., Ong, K.Y., Baranoski, J.M., Domestic Preparedness Program: testing of HAZMATCAD Detectors against chemical warfare agents summary report, Research and Technology Directorate, Soldier and Biological Chemical Command, Aberdeen Proving Ground, Maryland, February 2002.

Longworth, T.L., Ong, K.Y., Baranoski, J.M., Domestic Preparedness Program: testing of the Vaportracer against chemical warfare agents summary report, Research and Technology Directorate, Soldier and Biological Chemical Command, Aberdeen Proving Ground, Maryland, May 2002.

Longworth, T.L., Ong, K.Y., Johnson, M.A., Domestic Preparedness Program: testing of SAW Minicad Mkii Detector against chemical warfare agents summary report, Research Directorate, Soldier and Biological Chemical Command, Aberdeen Proving Ground, Maryland, September 2001.

Lucero, D.P., Performance characteristics of permeation tubes, *Anal. Chem.*, 43, 1744, 1971.

McNair, H.M. and Miller, J.M., *Basic gas chromatography*, John Wiley & Sons, New York, 1998.

Muse, W.T. and Buettner, L.C., Generation, sampling and analysis of GB vapor for inhalation toxicology studies, ECBC Technical Report, U.S. Army Edgewood Chemical and Biological Center, Edgewood, Maryland, in press.

O'Keeffe, A.E. and Ortman, G.C., Primary standards for trace gas analysis, *Anal. Chem.*, 38, 760, 1966.

Ong, K.Y., Multi-purpose chemical agent vapor generation system, ERDEC–TR-424, U.S. Army Edgewood Research, Development and Engineering Center (ERDEC), Aberdeen Proving Ground, Maryland, July 1997.

Ong, K.Y., Controlled multi-purpose chemical agent vapor generator system, U.S. Patent no. 5,728,927, March 17, 1998.

Ong, K.Y., Cajigas, J.C., Barnhouse, J.L., Analytical methodology for quantitative determination of O-ethyl S-(2-diisopropylaminoethyl) methylphosphonothiolate (VX), ERDEC–TR-424, U.S. Army Edgewood Research, Development and Engineering Center (ERDEC), Aberdeen Proving Ground, Maryland, ERDEC–TR-424, March 1998.

Ong, K.Y., Longworth, T.L., Barnhouse, J.L., Domestic Preparedness Program: testing of MSA Detector Tubes against chemical warfare agents summary report, Engineering Directorate, Soldier and Biological Chemical Command, Aberdeen Proving Ground, Maryland, July 2000.

Ong, K.Y., Cajigas, J.C., Barnhouse, J.L., Analytical methodology for quantitative determination of O-ethyl S-(2-diisopropylaminoethyl) methylphosphonothiolate (VX), ERDEC-TR-476, U.S. Army Edgewood Research, Development and Engineering Center (ERDEC), Aberdeen Proving Ground, Maryland, March 1998.

Ong, K.Y., Longworth, T.L., Barnhouse, J.L., Domestic Preparedness Program: testing of APD2000 Chemical Warfare Agent Detector against chemical warfare agents summary report, Research Directorate, Soldier and Biological Chemical Command, Aberdeen Proving Ground, Maryland, August 2000.

Mioduszewski, R.J., Reutter, S.A., Miller, L.L., Evaluation of airborne exposure limits for G-Agents: occupational and general population exposure criteria, ERDEC-TR-489, U.S. Army Edgewood Research, Development and Engineering Center (ERDEC), Aberdeen Proving Ground, Maryland, April 1998 (and February 2000 errata summary).

National Advisory Committee, National Advisory Committee for acute exposure guideline levels (AEGLs) for hazardous substances, proposed values, *Federal Register*, March 15, 2000, pp. 14186–14197.

Reutter, S.A., Mioduszewski, R.J., Thomson, S.A., Evaluation of airborne exposure limits for VX: occupational and general population exposure criteria, ECBC-TR-074, U.S. Army Edgewood Chemical and Biological Center, Maryland, February 2000.

Robards, K., Haddad, P.R., Jackson, P.E., *Principles and practice of modern chromatographic methods*, Academic Press, Harcourt Brace & Company, New York, 1994.

Skoog, A.S., Holler, F.J., Nieman, T.A., *Principles of Instrumental Analysis*, 5th ed., Harcourt Brace, Orlando, 1998.

Smart, J.K., History of chemical and biological detectors, alarms, and warning systems, U.S. Army Soldier and Biological Chemical Command, Aberdeen Proving Ground, Maryland, internal memo.

Tevault, D.E., Ong, K.Y., Wasserman, M.B., Vapor generation methods for chemical warfare agents, U.S. Army Edgewood Chemical and Biological Center, Maryland, ECBC-TR-148, March 2001.

Tomkins, B.A., Sega, G.A., Ho, C.H., Determination of Lewisite oxide in soil using solid-phase microextraction followed by gas chromatography with flame photometric or mass-spectrometric detection, *J. Chromatogr. A*, 909, 13–28, 2001.

Tuovinen, K., Paakkanen, H., Hanninen, O., Determination of soman and VX degradation products by an aspiration ion mobility spectrometry, *Anal. Chim. Acta*, 440, 151–159, 2001.

U.S. Department of the Army, ATCD-N, MEMORANDUM, Subject: Joint Service Operational Requirement (JSOR) for the Automatic Chemical Agent Detector Alarm (ACADA), Headquarters U.S. Army Training and Doctrine Command, Fort Monroe, Virginia, August 3, 1990.

U.S. Department of the Army, U.S. Army field manual 3-9, Potential military chemical/biological agents and compounds, Department of the Army, Washington, DC, December 12, 1990.

U.S. Department of the Army, Field manual 100-14, Risk management, Headquarters Department of the Army, Washington, DC, April 23, 1998.

U.S. Department of the Army, Army Regulation 385-61, The Army Chemical Agent Safety Program & PAM 385-61, Toxic chemical agent safety standards, Department of the Army, Washington, DC, March 27, 2002.

U.S. Department of Defense, Dictionary of military and associated terms, Joint Publication 1-02, April 12, 2001 (as amended through September 5, 2003).

U.S. Department of Health and Human Services, Documentation for immediately dangerous to life or health concentrations (IDLHs), U.S. Department of Health and Human Services, Public Health Service, Centers for Disease Control and Prevention, National Institute for Occupational Safety and Health, Washington, DC, May 1994.

U.S. Department of Justice, Guide for the selection of chemical agent and toxic industrial material detection equipment for emergency first responders, NIJ Guide 100-00, Vol. I, Washington, DC, June 2000.

Utriainen, M., Karpanoja, E., Paakkanen, H., Combining miniaturized ion mobility spectrometer and metal oxide gas sensor for the fast detection of toxic chemical vapours, *Sensors Actuators B*, B93, 17–24, 2003.

Index

A

B

C

D

E

F

G

H

J

K

L

M

N

S

T

U

V

W

9 781566 706681

For Product Safety Concerns and Information please contact our EU representative GPSR@taylorandfrancis.com
Taylor & Francis Verlag GmbH, Kaufingerstraße 24, 80331 München, Germany

www.ingramcontent.com/pod-product-compliance
Ingram Content Group UK Ltd.
Pitfield, Milton Keynes, MK11 3LW, UK
UKHW010956180926
475701UK00004B/397
* 9 7 8 0 3 6 7 3 9 4 0 2 8 *